P9-DXH-816

FUNDAMENTALS OF ELEMENTARY PARTICLE PHYSICS

McGraw-Hill series in fundamentals of physics: an undergraduate textbook program

INTRODUCTORY TEXTS

Young • Fundamentals of Mechanics and Heat
Kip • Fundamentals of Electricity and Magnetism
Young • Fundamentals of Optics and Modern Physics
Beiser • Concepts of Modern Physics

UPPER-DIVISION TEXTS

Barger and Olsson • Classical Mechanics: A Modern Perspective
Beiser • Perspectives of Modern Physics
Cohen • Concepts of Nuclear Physics
Elmore and Heald • Physics of Waves
Kraut • Fundamentals of Mathematical Physics
Longo • Fundamentals of Elementary Particle Physics
Meyerhof • Elements of Nuclear Physics
Reif • Fundamentals of Statistical and Thermal Physics
Tralli and Pomilla • Atomic Theory: An Introduction to Wave Mechanics

FUNDAMENTALS OF ELEMENTARY PARTICLE PHYSICS

Michael J. Longo
Professor of Physics
University of Michigan

McGraw-Hill Book Company
New York St. Louis San Francisco Düsseldorf
Johannesburg Kuala Lumpur London
Mexico Montreal New Delhi Panama
Rio de Janeiro Singapore Sydney Toronto

FUNDAMENTALS OF ELEMENTARY PARTICLE PHYSICS

1 2 3 4 5 6 7 8 9 0 KPKP 7 9 8 7 6 5 4 3 2

This book was set in Modern by The Maple Press Company. The editors were Jack L. Farnsworth and Carol First; the designer was Jo Jones; and the production supervisor was Thomas J. LoPinto. The drawings were done by Danmark & Michaels, Inc. The printer and binder was Kingsport Press, Inc.

Library of Congress Cataloging in Publication Data

Longo, Michael J 1935–
Fundamentals of elementary particle physics.

(McGraw-Hill series in fundamentals of physics)
Bibliography: p.
1. Particles (Nuclear physics) I. Title.
II. Series.
QC721.L8 539.7'21 71-39635
ISBN 0-07-038689-7

CONTENTS

PREFACE

The study of elementary particles is one of the most important areas of basic research in physics today. Nevertheless the average physics student, upon completion of his undergraduate courses, rarely has more than a vague notion of what the study of elementary particles is all about. It was for this reason that several years ago an undergraduate-level course on elementary particle physics was introduced at the University of Michigan. When I began to organize this course, it became immediately apparent that no appropriate text for such a course existed. There are perhaps dozens of books on elementary particle physics at the graduate and professional levels. There are also several books at what I might call the *Scientific American* level, but nothing at all between these two.

This book was written to try to bridge the gap. It is aimed primarily at the junior-senior level, but with augmentation it could be used for an introductory graduate course. I have tried to resist the temptation to write a book that my colleagues specializing in the field would find to their satisfaction. I shall be gratified by any insight they might gain from it, but this is incidental to the main purpose of the book.

The book is intended for a one-semester course in elementary particle physics. To keep within this stringent bound, I have deliberately avoided a historical approach. The spectrum of elementary particles is introduced in the first chapter, and their interactions are briefly discussed. Succeeding sections deal with the interactions between the particles in greater detail. A knowledge of the basics of quantum mechanics is assumed. This should be provided by the usual undergraduate quantum course taken either previously or concurrently. More advanced concepts such as scattering and symmetry operations, which are usually not covered in an introductory course on quantum mechanics, are presented in the book following an intuitive approach. Feynman-type diagrams are also introduced. Some knowledge of nuclear physics is supposed, but this could easily be gotten outside of a formal course.

While the book is basically written from an experimentalist's point of view, only a brief discussion of experimental technique is given (Chap. 2). This was meant only as a short introduction to the topic, to be supplemented by outside reading. Several good references for this purpose are listed in the Bibliography. In the original course upon which the book is based, several lectures were devoted to the discussion of recent experiments. The students were also assigned papers in which they were to discuss how they might perform an experiment to investigate a problem of current interest. These assignments served the very important functions of giving the students an idea of what experimental high-energy physics is really like, and of introducing them to the literature of the field.

This book is, of necessity, incomplete. In trying to condense a field of the breadth and complexity of elementary particles into a size suitable for a one-semester course, an author must be ruthless in picking and choosing material. One casualty, as mentioned above, is an adequate description of experimental techniques and particle accelerators (for which a number of references at a suitable level already exist). Another result is the lack of an in-depth discussion of certain theoretical topics of great current interest. I have tried to include material which would provide the student with a sampling of many aspects of the field and acquaint him with the basic concepts and terminology. The choice of material is largely a matter of taste, and I apologize to my colleagues for many sins of omission. In general I have assumed that the text will be reinforced by outside reading; this allows the instructor opportunity to introduce material of his own choosing.

It is a great pleasure to acknowledge the many people who contributed to this book. Particular thanks go to Robert T. Deck, an old friend and colleague, for his diligent and patient attempts to correct my many lapses in both grammar and physics, to Lawrence W. Jones for suggestions on problems and other material, and to Mrs. LaVaughn Karapostoles for her considerable skill in typing the manuscript. I am especially grateful to the Particle Data

Group at the Lawrence Berkeley Laboratory for providing figures and tables used in the text, in particular Fig. 3-17, Appendix A, and Table B-1; their compilations were also the source of data for many other figures and tables.

Michael J. Longo

FUNDAMENTALS OF ELEMENTARY PARTICLE PHYSICS

CHAPTER 1

introduction

It seems probable to me, that God in the Beginning form'd Matter in solid, massy, hard, impenetrable, moveable Particles, of such Sizes and Figures, and with such other Properties, and in such Proportion to Space, as most conduced to the End for which he form'd them.

From Sir Isaac Newton,
"Opticks," Dover, New York, 1952

The modern physicist working in elementary particle research is continuing a quest for understanding of the basic structure of matter which has gone on for thousands of years. About 400 B.C. Democritus postulated that matter was built up out of indivisible and invisible particles which were given the name *atomoi*. For over 20 centuries no significant progress was made in searching for such particles. It was not until 1897 that the first of the particles we now would call "elementary" was identified by J. J. Thompson. He showed that "cathode rays," produced when high voltage was applied to electrodes in an evacuated vessel, were negatively charged particles with a charge-to-mass ratio over a thousand times as great as that for hydrogen ions. He later adopted the name *electrons* for these particles.[1]

1-1 WHAT IS AN ELEMENTARY PARTICLE?

It would seem reasonable to start off a text on elementary particles by defining what an elementary particle is. However, the historical concept of elementary particles as the ultimate, indivisible building blocks of matter seems to have lost its relevance as men have probed deeper into the structure of matter. We now know that if two protons of sufficient energy collide, proton-antiproton pairs can be produced, as well as an indefinitely large number of "fragments" such as pions and kaons. Furthermore, the list of possible kinds of

[1] Though the history of elementary particle research is a fascinating topic in itself, we regrettably cannot discuss it here. A good reference is H. A. Borse and L. Motz, eds., "The World of the Atom," vols. I and II, Basic Books, New York, 1966.

fragments becomes longer with almost every issue of *The Physical Review.* It has also become apparent that all the particles are coupled, more or less strongly, to each other. The term "elementary particle" has therefore become unfashionable. However, for lack of a better term we shall continue to refer to anything less massive than a deuteron as an elementary particle or, more simply, a particle. The term "elementary particle physics" has also fallen into disfavor, but we shall continue to use this term as well as the more or less synonymous term "high-energy physics."

1-2 THE HIERARCHY OF INTERACTIONS BETWEEN PARTICLES

The interactions between particles appear to be divisible into four classes. In order of decreasing strength these are:

1. *Strong interactions.* These are most familiar as the forces between nucleons which bind nuclei together. These interactions have a characteristic range of approximately 10^{-13} cm. Not all particles participate in the strong interactions (e.g., the electron and muon do not). The particles that do are collectively referred to as *hadrons.*
2. *Electromagnetic interactions.* These are the familiar forces between *charged* particles. They are responsible for the binding of atoms and molecules. They are typically about 100 times weaker than the strong interactions. The electromagnetic interaction has a long range; however the effective range is limited because bulk matter tends to be electrically neutral. The electromagnetic interaction is mediated by the photon.
3. *Weak interactions.* These are responsible for β decay and related effects. Crudely speaking, they are about 10^{-13} times as strong as the strong interaction. Their range is $\ll 10^{-13}$ cm.
4. *Gravitational interaction.* These are by far the weakest of the interactions on the microscopic scale, typically about 10^{-40} times as strong as the strong interactions for separations comparable to nuclear dimensions. However the gravitational interaction has a long range and is charge independent. We therefore have the somewhat paradoxical result that gravitational interactions are insignificant on the atomic and nuclear

scale, but completely dominant on the cosmic scale. In what follows, we shall neglect the gravitational interaction completely.

Note that these classes are not exclusive. A proton, for example, participates in all four kinds of interactions. A neutrino, on the other hand, is only involved in weak and gravitational interactions. It is not clear whether this division of the interactions into four types should be considered a measure of our knowledge or of our ignorance. Eventually we should hope to find connections between the various types, but so far this has not been possible. There seems to be a relation between the mass of a particle and the strength of its interactions, with the hadrons being the most massive particles and also the most strongly interacting. As we shall discuss below, the stronger interactions seem to possess more "symmetry." All these facts are no doubt significant, but they are in no sense understood. These are just examples of the many unsolved problems in elementary particle physics which make it a frontier field par excellence.

The strong, electromagnetic, and weak interactions will be discussed in detail in following chapters.

1-3 THE "STABLE" PARTICLES[1]

We shall now list the more stable particles and their basic properties, the dramatis personae as it were. The significance of some of these properties will only become clear later on. Table 1-1, based on a compilation by the Particle Data Group,[2] lists the particles whose lifetimes are $\gtrsim 10^{-21}$ sec. This marks the boundary between particles whose decay takes place through a strong interaction (lifetimes $\lesssim 10^{-21}$ sec) and those decaying through a weak or electromagnetic interaction (lifetimes $\sim 10^{-10}$ sec and $\sim 10^{-20}$ sec respectively).

The table summarizes the results of decades of work by thousands

[1] The succeeding sections of this chapter, of necessity, introduce many facts and ideas. The beginning student should try to become generally familiar with the material in this chapter the first time around, and then return to it in conjunction with the study of Chap. 3.

[2] This group provides an important service by collecting, evaluating, and summarizing data on elementary particles. A complete compilation can be found in a supplement to *Rev. Mod. Phys.* **43** (April, 1971).

of scientists. A complete description of the experimental techniques employed in determining these data would take volumes. The reader is urged to consult some of the references in the Bibliography to gain some insight into the experimental work involved in their determination.

The first two columns in Table 1-1 give the common name and standard symbol for the particle. The next three columns give its quantum numbers (where applicable). S is the so-called *strangeness;* T is the *isotopic spin;* G is the G *parity;* J is the *spin*, or *total angular momentum;* and P is the *parity*. These will be discussed in the following sections.

The next column in the table lists the mass.[1] The remaining columns give the lifetime and principal decay modes.

The elementary particles, excepting the photon γ, can be divided into three major groups:

1. *Leptons* ("light" particles). These include the neutrino ν, the electron e, and the muon μ. As a group, the leptons are characterized by the fact that they do *not* participate in the strong interactions. All the leptons have spin $\frac{1}{2}$.
2. *Mesons.* These are particles of intermediate mass and include the pion π, kaon K, and η. All are unstable, decaying via weak or electromagnetic interactions. However, all also undergo strong interactions. All mesons have zero or integral spin. In addition to those listed, there are many shorter-lived mesons which decay through strong interactions.
3. *Baryons* ("heavy" particles). These include the nucleons n and p and the more massive particles Λ, Σ, Ξ, and Ω. (The last four are generically called *hyperons.*) The baryons interact strongly. All have spin $\frac{1}{2}$ or $\frac{3}{2}$. In addition to the baryons listed, there are a large number of shorter-lived states that have masses greater than the nucleon mass and decay by strong interactions.

The photon seems to be in a class by itself. It is, of course, stable and massless. It has a spin 1 and apparently participates (directly) only in the electromagnetic and gravitational interactions.

[1] We shall conform to common usage and measure masses in million electron volts: 1 MeV = 10^6 eV; 1 GeV (or BeV) = 10^9 eV. Conversion factors and other useful constants are given in Appendix A.

Table 1-1 The "stable" particles

Common name	Symbol	S	T^G	J^P	Mass, MeV	Mean life τ, sec	Decay modes: Mode	Decay modes: Fraction, %
Photon	γ	0	$T = 0$ or $T = 1$	1^-	$0\ (<2 \times 10^{-21})$	Stable		
Leptons								
Neutrino	ν_e	0		$\frac{1}{2}$	$0\ (<60\ \text{eV})$	Stable		
	ν_μ				$0\ (<1.0)$			
Electron	e^-	0		$\frac{1}{2}$	0.51100	Stable ($>2 \times 10^{21}$ yr)		
Muon	μ^-	0		$\frac{1}{2}$	105.660	2.198×10^{-6}	$e\nu\bar{\nu}$	100
							$e\gamma$	$(<2 \times 10^{-6})$
Mesons								
Pion	π^+	0	1^-	0^-	139.58	2.602×10^{-8}	$\mu\nu$	≈ 100
							$e\nu$	1.24×10^{-2}
							$\pi^0 e\nu$	1.0×10^{-6}
	π^0	0	1^-	0^-	134.97	0.8×10^{-16}	$\gamma\gamma$	98.8
							$\gamma e^+ e^-$	1.16
							$\gamma\gamma\gamma$	$<5 \times 10^{-4}$
Kaon	K^+	+1	$\frac{1}{2}$	0^-	493.8	1.237×10^{-8}	$\mu\nu$	63.8
							$\pi^+\pi^0$	20.9
							$\pi\pi^+\pi^-$	5.6
							$\pi\pi^0\pi^0$	1.7
							$\mu\pi^0\nu$	3.2
							$e\pi^0\nu$	4.9
							$\pi\pi^\mp e^\pm\nu$	3.3×10^{-3}
							$\pi\pi^\mp\mu^\pm\nu$	0.9×10^{-3}
							$e\nu$	1.3×10^{-3}
	K^0	+1	$\frac{1}{2}$	0^-	497.8		$K^0 \approx \frac{1}{2}(K_S{}^0 + K_L{}^0)$	
	$(K_S{}^0)$			0^-	$m(K_L{}^0) - m(K_S{}^0) \approx 0.464\hbar/\tau_S$	0.86×10^{-10}	$\pi^+\pi^-$	68.7
							$\pi^0\pi^0$	31.3
							$\mu^+\mu^-$	$<0.7 \times 10^{-3}$
							$\pi^+\pi^-\gamma$	2×10^{-1}

	$(K_L{}^0)$			0^-		5.17×10^{-8}	$\pi^0\pi^0\pi^0$	21.4
							$\pi^+\pi^-\pi^0$	12.6
							$\pi\mu\nu$	26.8
							$\pi e\nu$	38.9
							$\pi^+\pi^-$	0.16
							$\pi^0\pi^0$	0.09
							$\gamma\gamma$	5×10^{-2}
							$\mu^+\mu^-$	$<1.9 \times 10^{-7}$
							e^+e^-	$<1.6 \times 10^{-7}$
(Eta)	η^0	0	0^+	0^-	548.8	$\Gamma \approx 2.6$ keV ($\Gamma = \hbar/\tau$)	$\gamma\gamma$	38.6
							$\pi^0\gamma\gamma$	3.3
							$3\pi^0$	30
							$\pi^+\pi^-\pi^0$	23
							$\pi^+\pi^-\gamma$	4.7
Baryons								
Proton	p	0	$\frac{1}{2}$	$\frac{1}{2}^+$	938.259	Stable ($>2 \times 10^{28}$ yr)		
Neutron	n	0	$\frac{1}{2}$	$\frac{1}{2}^+$	939.553	0.93×10^3	$pe^-\bar{\nu}$	100
(Lambda)	Λ^0	−1	0	$\frac{1}{2}^+$	1,115.6	2.5×10^{-10}	$p\pi^-$	64
							$n\pi^0$	36
							$pe\nu$	0.8×10^{-1}
							$p\mu\nu$	1.4×10^{-2}
(Sigma)	Σ^+	−1	1	$\frac{1}{2}^+$	1,189.4	0.80×10^{-10}	$p\pi^0$	52
							$n\pi^+$	48
							$p\gamma$	1.2×10^{-1}
	Σ^0	−1	1	$\frac{1}{2}^+$	1,192.5	$<10^{-14}$	$\Lambda\gamma$	100
	Σ^-	−1	1	$\frac{1}{2}^+$	1,197.4	1.49×10^{-10}	$n\pi^-$	100
							$ne^-\nu$	1.1×10^{-1}
							$n\mu^+\nu$	0.45×10^{-1}
							$\Lambda e^-\nu$	0.6×10^{-2}
(Xi) or "cascade"	Ξ^0	−2	$\frac{1}{2}$	$\frac{1}{2}^+$	1,315	3.0×10^{-10}	$\Lambda\pi^0$	100
							$p\pi^-$	$<0.9 \times 10^{-1}$
	Ξ^-	−2	$\frac{1}{2}$	$\frac{1}{2}^+$	1,321.3	1.66×10^{-10}	$\Lambda\pi^-$	100
							$\Lambda e^-\nu$	0.7×10^{-1}
							$\Sigma^0 e^-\nu$	$<0.5 \times 10^{-1}$
							$n\pi^-$	$<1.1 \times 10^{-1}$
(Omega)	Ω^-	−3	0	$\frac{3}{2}^+$?	1,672.5	1.3×10^{-10}	$\Xi^0\pi^-$	(world sample ~30 events)
							$\Xi^-\pi^0$	
							ΛK^-	

1-4 PARTICLES AND ANTIPARTICLES

All particles so far discovered have their counterpart *antiparticles*. The most familiar example is the positron, which is the "anti-electron." The antiparticle has the same mass, spin, and lifetime as the corresponding particle, but opposite quantum numbers as listed below:

Charge. Equal but opposite in sign
Strangeness (defined later). Equal but opposite in sign
Magnetic moment. Equal but opposite in sign (relative to spin direction)
Decay schemes. Complementary, for example,

$$n \rightarrow p + e^- + \bar{\nu} \qquad \bar{n} \rightarrow \bar{p} + e^+ + \nu$$

An antiparticle is commonly designated by the particle symbol with a bar over it, though the bar is often omitted where it is redundant (for example, e^+ and e^-, and π^+ and π^-). The antiphoton and anti-π^0 meson by these rules have the same quantum numbers as the γ and π^0 respectively, and in these two cases the particle and antiparticle are identical. In other words, $\bar{\gamma} = \gamma$ and $\bar{\pi}^0 = \pi^0$.

To some extent the question of which is the particle and which is the antiparticle is one of definition. Generally speaking, in our corner of the universe the particles are more abundant or easier to create than antiparticles, but this should not be thought of as a fundamental distinction.

1-5 CONSERVATION OF LEPTONS AND BARYONS

In view of the fact that most of the elementary particles are unstable, it comes as something of a surprise that the proton is stable against decay into pions or light particles. In fact, as indicated in Table 1-1, the experimental limit on the lifetime of the proton is $>10^{28}$ years, or about 18 orders of magnitude greater than the estimated age of the universe. The absence of such decay modes for nucleons and of similar decay modes for hyperons is embodied in the *law of conservation of baryon number*. This states that if we assign a *baryon number* B of $+1$ to each nucleon or hyperon and -1 to an antinucleon or antihyperon, then in a closed system

$$\sum B = \text{constant}$$

A similar conservation law for leptons is also observed.[1] For example, the decay $\mu^+ \rightarrow e^+ + \nu$ does not occur. Instead, the muon decays by $\mu^+ \rightarrow e^+ + \nu + \bar{\nu}$. We therefore invoke a conservation law for leptons. If μ^-, e^-, and ν are assigned a *lepton number* L of $+1$ and μ^+, e^+, and $\bar{\nu}$ are assigned $L = -1$, then for a closed system,

$$\sum L = \text{constant}$$

These conservation laws, in contrast to other conservation laws which we shall discuss shortly, are absolute and appear to hold under all circumstances. It is important to note that a similar conservation law for mesons and photons does not exist. A lamp filament, for example, can create photons in arbitrarily large numbers.

1-6 FERMIONS VERSUS BOSONS

In quantum mechanics a system of identical particles 1, 2, 3, . . . is described by a wave function $\psi(\mathbf{r}_1,\mathbf{r}_2,\mathbf{r}_3, \ldots ;\mathbf{s}_1,\mathbf{s}_2,\mathbf{s}_3, \ldots)$, where $\mathbf{r}_i$ is the coordinate vector, and $\mathbf{s}_i$ the spin of the ith particle. It can be shown that the wave function must be either symmetrical (even) or antisymmetrical (odd) with respect to the interchange of coordinates of any pair of identical particles.[2] If it is symmetrical, the particles are called *bosons;* if it is antisymmetrical, they are called *fermions.* This symmetry or antisymmetry of wave functions has far-reaching consequences. An antisymmetrical wave function must vanish as two identical particles approach each other. As a

[1] The law of conservation of leptons, as given here, will have to be generalized slightly to take into account the existence of two kinds of neutrinos. This will be discussed in Chap. 6.

[2] L. I. Schiff, "Quantum Mechanics," 3d ed., chap. 10, McGraw-Hill, New York, 1968.

Enrico Fermi. Born 1901 in Rome; died 1954. His early work included discovery of the so-called Fermi-Dirac statistics for particles with half-integer spin (fermions). While at the University of Rome, he and his group carried out a series of classic experiments to study artificial radioactivity induced by neutron bombardment for which he later received the Nobel Prize. The theory of β decay that he proposed in 1934 was the basis for the current theory of weak interactions. In 1939 he emigrated to the United States. His contributions to the development of the first chain-reacting pile during the war are well known. After the war he and his group at the University of Chicago started a series of experiments to study pion-nucleon scattering with pion beams from the Chicago cyclotron. Fermi also made significant contributions to astrophysics. He was one of the very few physicists who have done outstanding work in both theory and experiment. He was a superb teacher and writer; his books and papers are noted for their clarity. Element 100 in the periodic table is named in his honor. (*Photograph from the Niels Bohr Library, American Institute of Physics.*)

result, two fermions in the same quantum state exhibit a strong mutual repulsion, which leads to the Pauli exclusion principle. On the other hand, no restriction exists on the number of ideal (noninteracting) bosons that can be assembled at a point.[1]

Empirically it is found that particles with half-integer spins $\frac{1}{2}$, $\frac{3}{2}$, $\frac{5}{2}$, . . . are fermions, while particles with integer spins are bosons. Thus leptons and baryons are fermions; mesons and photons are bosons.

A system whose orbital angular-momentum quantum number ℓ is even, 0, 2, . . . , has the spatial part of its wave function symmetrical, or even. Two identical, spinless bosons (for example, π^0 mesons) must therefore be in a state with ℓ even, so that the wave function is symmetrical. For particles with spin, the overall wave function which describes the system is the product of a spin part and a space part. For example, since protons are fermions, two protons with their spins parallel (i.e., symmetrical) must have ℓ odd, while two protons with spins antiparallel must be in a state with ℓ even.[2]

1-7 STRANGENESS

Another quantum number and related conservation law is that of *strangeness*. Its importance was suggested by experimental results which showed that the K mesons and hyperons could only be produced in pairs; for example,[3]

$$p + p \nrightarrow p + p + K^0$$

The experimental results can be systematized by assigning a new quantum number S which is nonzero only for the K meson and the

[1] It should be emphasized that real bosons do interact with each other. Mesons, for example, are bosons and interact between themselves through strong interactions.

[2] The Pauli principle can be generalized to any system of nucleons by considering the isospin (defined below) just like the ordinary spin. Then the overall wave function of the system is the product of three parts: a spin part, an isospin part, and a space part. For fermions the overall product must be odd. This is often referred to as the *generalized Pauli principle*.

[3] We use the symbol $\nrightarrow$ for a reaction that is not allowed or not observed.

hyperons. Thus K^0 has $S = +1$; $\bar{K}^0$ has $S = -1$. Since $\pi^- + p \rightarrow K^0 + \Lambda^0$ does occur, we assign $S = -1$ to the Λ^0. The pions and nucleons have $S = 0$.

We also know that K^0's can decay by $K^0 \rightarrow \pi^+ + \pi^-$. This obviously does not conserve strangeness. Since the decay rate is typical of that for a weak interaction, we conclude that the weak interactions do *not* conserve strangeness. The electromagnetic interactions, on the other hand, do appear to conserve S.

Strangeness assignments for the particles are given in Table 1-1. As previously noted, the antiparticles have equal and opposite strangeness. The strangeness of the leptons is undefined; since strangeness is not conserved in the weak decay modes which connect the hadrons to the leptons, a unique strangeness cannot be assigned to the leptons.

1-8 ISOTOPIC SPIN

From Table 1-1 it is apparent that the particles are grouped into families or multiplets whose masses are very nearly equal (e.g., the π^0, $\pi^\pm$). This familial relationship has of course been recognized in naming the particles. It is therefore convenient to treat the members of a multiplet as different states of the same particle. The states (members) are distinguished by a quantum number, *isotopic spin* or *isospin*, which is completely analogous to ordinary spin. Just as the spin quantum number determines the orientation of the spin vector in real space, the isotopic spin quantum number can be thought of as determining the orientation of the isospin vector in a fictitious isospin space.[1] For a particle of spin s, there are $2s + 1$ possible orientations of the spin vector relative to a given direction; similarly, for a multiplet with isospin T, there are $2T + 1$ members of the multiplet. For example, we can accommodate the three pions by assigning $T = 1$ to the pion multiplet. Since the particles in each multiplet are also distinguished by their charge, the orientation of $\mathbf{T}$ can be thought of as determining the charge. By convention

[1] It should be emphasized that isospin space has no physical significance except as a convenient way of relating the algebra of isotopic spin to that of ordinary spin.

the component of **T** along the Z axis in isospin space is related to the charge Q by

$$T_3 = \frac{Q}{e} - \frac{B}{2} - \frac{S}{2} \tag{1-1}$$

where B is the baryon number ($+1$ for a nucleon or hyperon, -1 for an antinucleon or antihyperon), and S is the strangeness. The other two components of **T** are then undefined (as they would be for ordinary spin vectors in quantum mechanics) and have no special physical significance.

The reason for going through the bother of defining the quantum number T is that experimentally it is found that the strong interactions conserve isospin. In other words, the total isospin of the final state of a strong reaction must be the same as that of the initial state. Thus some reactions are forbidden by isospin conservation. The electromagnetic and weak interactions, on the other hand, do *not* conserve isospin, so a "forbidden" reaction might still go by one of these, but at a much lower rate than it would through strong interactions.[1] The isotopic spin of the leptons is not defined, as the leptons do not participate in strong interactions and isotopic spin is not conserved in weak reactions.

As a result of isospin conservation, the properties of a system in regard to the strong interactions are determined only by its total isospin T independently of T_3 (or charge). Thus the π^+-π^+ system ($T_3 = +2$, $T = 2$) is equivalent as far as the strong interactions are concerned to the π^--π^- system ($T_3 = -2$, $T = 2$). The π^+-n system ($T_3 = \frac{1}{2}$, $T = \frac{1}{2}$ or $\frac{3}{2}$) can be treated as a mixture of $T = \frac{1}{2}$ and $T = \frac{3}{2}$. Furthermore the π^--p system can be shown to be the same mixture of $T = \frac{1}{2}$ and $T = \frac{3}{2}$ as the π^+-n system, so that the two systems are equivalent as far as their strong interactions are concerned.

Table 1-2 summarizes the isotopic spin assignments of some of the multiplets. (For each multiplet there is a corresponding multiplet with the antiparticles.)

[1] An example is the decay of the η^0, which is forbidden to decay into three pions via a strong interaction (Sec. 3-7), although the decay $\eta^0 \to 3\pi$ is still observed with a rate characteristic of an electromagnetic interaction.

Table 1–2 Some isotopic spin multiplets

Particle	*T*	*Multiplicity*	*States*	T_3	**T**
N	$\frac{1}{2}$	Doublet	p	$+\frac{1}{2}$	↑
			n	$-\frac{1}{2}$	↓
Λ	0	Singlet	Λ^0	0	
Σ	1	Triplet	Σ^+	+1	↑
			Σ^0	0	→
			Σ^-	−1	↓
Ω	0	Singlet	Ω^-	0	
π	1	Triplet	π^+	+1	↑
			π^0	0	→
			π^-	−1	↓
K	$\frac{1}{2}$	Doublet	K^+	$+\frac{1}{2}$	↑
			K^0	$-\frac{1}{2}$	↓

1-9 CHARGE CONJUGATION AND PARITY[1]

Charge conjugation C is the operation of changing a particle to its antiparticle; for example,

$$\mathsf{C}(\pi^+) = \pi^-$$
$$\mathsf{C}(p) = \bar{p}$$
$$\mathsf{C}(\pi^0) = \pi^0$$

Experimentally it has been found that the strong and electromagnetic interactions, but not the weak, are symmetrical (invariant) with respect to charge conjugation. Thus the strong and electromagnetic interactions between π^- and p are the same as those between π^+ and $\bar{p}$, and the reaction $\pi^- + p \rightarrow K^0 + \Lambda^0$ should proceed at the same rate as $\pi^+ + \bar{p} \rightarrow \bar{K}^0 + \bar{\Lambda}^0$.

The parity operation P is a reflection of all coordinates through the origin. Thus $\mathsf{P}(\mathbf{r}) = -\mathbf{r}$, where $\mathbf{r}$ is any radius vector. Invariance under P would imply that nature is symmetrical with respect to mirror imaging, which for a mirror in the XY plane, is equivalent to the operation $z \rightarrow -z$. This symmetry is obviously broken locally

[1] If desired, the study of this section can be postponed until Chap. 6. A slight familiarity with the concepts is all that is required at this stage.

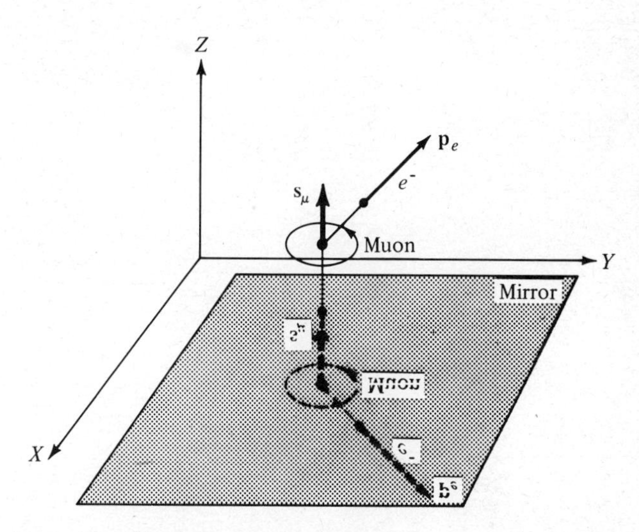

Figure 1-1 The decay of a polarized muon, $\mu \rightarrow e +$ neutrinos. The mirror image of a muon with spin up also has spin up since the sense of rotation is the same for the mirror image. However the momentum vector of the electron (formed when the muon decays) is inverted. If, in a large sample of decays, the electrons exhibit a preference to go up or down relative to the muon spin, symmetry with respect to mirror imaging does not hold.

on a macroscopic scale (e.g., there are more right-handed people than left-handed); however, if we could sample a large number of "universes," we might expect symmetry. Similarly, on the submicroscopic scale we might expect that there is no preferred handedness.

To be specific, let us consider the decay of a sample of polarized muons, i.e., muons with their spin **s** aligned along, say, the Z axis as shown in Fig. 1-1. For a mirror in the XY plane,[1] the mirror image of the situation shown is a muon whose spin is still along the $+Z$ axis (since the sense of rotation of the muon is the same in the mirror), but the decay electron is going generally downward. Thus if the decay electrons tended to come off preferentially parallel to the muon spin, the expected symmetry between the system and its mirror image would not hold; in the mirror the electrons come off anti-

[1] The argument is true for any mirror; however the details of the analysis will depend on the location of the mirror.

parallel to the muon spin, so that the real situation is distinguishable from its mirror image.

In other words, symmetry with respect to the parity operation implies that the distribution of electrons emitted in the decay of muons polarized along the Z axis must be symmetrical about the XY plane. In fact, this is found *not* to be true, so we say that μ decay, and weak interactions in general, violate P invariance.

Note that the P operation changes the sign of coordinates and momenta

$$\mathsf{P}(\mathbf{r}) = -\mathbf{r}$$
$$\mathsf{P}(\mathbf{p}) = -\mathbf{p}$$

but does not affect spins and angular momenta

$$\mathsf{P}(\mathbf{s}) = \mathbf{s}$$
$$\mathsf{P}(\mathbf{L}) = \mathbf{L}$$

where $\mathbf{L} = \mathbf{r} \times \mathbf{p}$. The vectors $\mathbf{s}$ and $\mathbf{L}$ are examples of *pseudovectors,* or *axial* vectors. Experiments show that the strong and electromagnetic interactions are invariant with respect to parity, but large violations are found in weak interactions. We shall discuss C and P invariance further when we talk about the weak interactions in Chap. 6.

The elementary particles themselves (actually their wave functions) generally have a well-defined symmetry under P.[1] Thus it is found that

$$\mathsf{P}(p) = +p$$
$$\mathsf{P}(\pi) = -\pi$$
$$\mathsf{P}(K) = -K$$

Particles whose wave functions change sign under P are said to have *odd intrinsic parity.* For a system of two or more identical particles, the parity of the system is the product of the parity of the individual particles and that associated with the angular-momentum state. Thus a system of three π^0's must have ℓ odd so that the overall wave function is symmetrical (since π^0's are bosons).

Intrinsic parities are listed in Table 1-1 as a + or − superscript on the spin J. As with strangeness, none is assigned to the leptons.

[1] See Schiff, *op. cit.*, p. 226.

Table 1-3 Conservation laws and their applicability to the strong, electromagnetic, and weak interactions

	Type of interaction		
Quantity conserved	*Strong*	*em*	*Weak*
Energy-momentum	Yes	Yes	Yes
Angular momentum	Yes	Yes	Yes
Electric charge	Yes	Yes	Yes
Baryon number	Yes	Yes	Yes
Lepton number	Yes	Yes	Yes
Parity	Yes	Yes	No
Strangeness	Yes	Yes	No
Charge conjugation	Yes	Yes	No
Isospin	Yes	No	No

1-10 SUMMARY OF CONSERVATION LAWS

Table 1-3 summarizes our present understanding of the behavior of the strong, electromagnetic (em), and weak interactions in regard to the conservation laws and related symmetries. It is interesting to note the apparent correlation between the degree of symmetry and the strength of the various interactions. However the significance of this correlation, if any, is not yet known. Again we are presented with some tantalizing clues to one of nature's most profound secrets. For the time being we must view Table 1-3 as a collection of empirical facts in search of an explanation.

PROBLEMS

For a review of relativistic kinematics and kinematics tables, see Appendix B.

1-1 Which of the following are examples of strong, electromagnetic, weak, or gravitational interactions?
(*a*) Binding forces in the solar system
(*b*) Binding forces in molecules
(*c*) Binding forces in atoms
(*d*) Binding forces in nuclei

(*e*) The decay $K^0 \to \pi^+ + \pi^-$
(*f*) The decay $\pi^0 \to \gamma + \gamma$

1-2 Which of the following reactions conserve strangeness?
(*a*) $\pi^0 \to \gamma + \gamma$ (*b*) $K^0 \to \pi^+ + \pi^-$
(*c*) $\Sigma^0 \to \Lambda^0 + \gamma$ (*d*) $\bar{\Sigma}^+ \to \bar{n} + \pi^+$
(*e*) $\pi^- + p \to K^0 + \Lambda^0$ (*f*) $\pi^- + p \to \bar{K}^0 + \Lambda^0$
(*g*) $K^- + p \to K^0 + n$ (*h*) $p + p \to K^0 + p + p$
(*i*) $\eta^0 \to K^0 + \gamma$

1-3 Indicate whether the following reactions are most likely to proceed through a strong, electromagnetic, or weak interaction. If a reaction is absolutely forbidden, indicate why.
(*a*) $n \to p + e^- + \bar{\nu}$ (*b*) $\Lambda^0 \to \Sigma^0 + \gamma$
(*c*) $p + p \to \pi^+ + d$ (*d*) $n + p \to \gamma + d$
(d = deuteron)
(*e*) $\pi^- + p \to \pi^0 + n$ (*f*) $K^+ \to \mu^+ + \nu$
(*g*) $\mu^+ \to e^+ + \nu$ (*h*) $\Omega^- \to K^- + \bar{K}^0$
(*i*) $\pi^- \to \mu^- + \bar{\nu}$ (*j*) $\bar{p} + p \to e^+ + e^-$
(*k*) $n \to e^- + e^+$

1-4 Calculate the mean free path for decay of a 3 GeV/c π^+ beam.[1] Repeat for a 3 GeV/c π^0 beam.

1-5 Calculate the muon energy in the decay $K^+ \to \mu^+ + \nu$ in the rest frame of the K^+.

1-6 Calculate the maximum electron energy in the decay $\mu^+ \to e^+ + \nu + \bar{\nu}$. (What configuration of decay angles gives the maximum energy e^+?)

1-7 According to Table 1-1 the free neutron has a half-life of approximately 15 min for the decay $n \to p + e^- + \bar{\nu}$. Why are neutrons that are bound in nuclei generally stable?

[1] The unit MeV/c for momentum may seem strange to those unfamiliar with it. A momentum of 1 MeV/c corresponds to pc = 1 MeV.

CHAPTER 2

the tools of elementary particle research

In this chapter we shall briefly discuss the experimental aspects of elementary particle research. This is a subject about which volumes have been written. We shall only give a brief introduction to the topic. The reader is urged to consult one of the many books on experimental techniques, some of which are listed in the Bibliography.

2-1 ACCELERATORS

The experimental study of elementary particles concerns, first of all, their production. In the early days of elementary particle research, radioactive sources and cosmic radiation were important sources of high-energy particles. After the construction of the first cyclotron by E. O. Lawrence and M. S. Livingston in 1931, higher- and higher-energy accelerators were rapidly developed. The construction of the first accelerators in the billion-electron-volt range early in the 1950s began a new and exciting period in elementary particle research.[1] Nowadays almost all experiments are done with beams of particles from accelerators. Cosmic rays, while an important subject of research in their own right, cannot compete with accelerator beams in intensity and controllability. The primary cosmic-ray spectrum, however, does contain appreciable numbers of particles with energies greater than 1,000 GeV, well beyond the range of existing accelerators. Experiments which utilize these, though

[1] The first machine capable of accelerating protons to energies >1 GeV was the 3-GeV Cosmotron at Brookhaven National Laboratory (completed in 1952 and no longer operating). The 6-GeV Bevatron at the Lawrence Radiation (now Lawrence Berkeley) Laboratory was completed about 2 years later.

E. O. Lawrence (right) with M. S. Livingston. Lawrence was born in 1901 in South Dakota and died in 1958. He received the Nobel Prize in 1939 for the discovery and development of the cyclotron. Livingston assisted in the development of the cyclotron, first as a graduate student under Lawrence's direction and later as an instructor at the University of California. Their paper in 1932, in which they discussed the theory and operation of the cyclotron, marked the beginning of a new era in the study of elementary particles. The photograph shows Lawrence and Livingston with their "huge" $27\frac{1}{2}$-in. cyclotron, which in 1933 reached 4 MeV. The 80-ton magnet for the cyclotron was built from part of an obsolete radio transmitter.

During World War II Lawrence directed an effort to separate U^{235} from U^{238} by mass-spectroscopic techniques. He also participated in the discovery of many radioactive isotopes of known elements and pioneered in the application of the cyclotron to medical and biological problems. (*Photograph from the Lawrence Berkeley Laboratory.*)

Table 2-1 High-energy accelerators (>2 GeV)

Location	*Name*	*Energy,* GeV*	*Intensity, 10^{12} particles/pulse*	*Max. rep. rate, sec^{-1}*
I. Proton accelerators				
LBL, Berkeley, Calif.	Bevatron	6.2	5	0.2
ANL, Argonne, Ill.	ZGS	12.5	3	0.4
BNL, Upton, N.Y.	AGS	33	10	1.0
NAL, Batavia, Ill.	(Synchrotron)	200–500	~50†	0.35–0.12
ITEP, Moscow, USSR	(Synchrotron)	7.5	0.6	0.25
JINR, Dubna, USSR	(Synchrotron)	10	0.1	0.1
IHEP, Serpukhov, USSR	(Synchrotron)	76	1	0.1
CERN, Geneva, Switzerland	PS	28	2	0.40
RHEL, Harwell, England	Nimrod	8	3.5	0.5
CENS, Saclay, France	Saturne	3	1.2	0.33
II. Electron accelerators				
CEA, Cambridge, Mass.	CEA	6	0.1	60
Cornell Univ., Ithaca, N.Y.	(Synchrotron)	12	0.1	60
SLAC, Stanford, Calif.	SLAC	21	0.8	360
Phys. Inst., Bonn, Germany		2.5	0.06	50
DESY, Hamburg, Germany	DESY	7.5	0.6	50
Lab. de l'Accélérateur Linéaire, Orsay, France		2.3		
NPL, Daresbury, England	NINA	5	0.2	50
Phys. Tech. Inst., Kharkov, USSR	LU-2	2	0.15	50
Inst. Phys., Yerevan, USSR		6	0.1	50

* Maximum kinetic energy.

† Design parameter. Not yet in routine operation.

generally limited in statistical accuracy, provide important insights into the ultrahigh-energy domain.

Table 2-1 lists the major high-energy accelerators in the world and their important characteristics.[1] The highest-energy proton

[1] A discussion of the theory or operation of accelerators is beyond the scope of this book. For references consult the Bibliography.

Figure 2-1 An aerial view of the 200 to 500 GeV proton synchrotron near Batavia, Illinois. The diameter of the ring is 2 km (1.24 miles). The smaller ring in the right foreground is the 8-GeV "booster" which injects 8-GeV protons into the main ring. (*Photograph from the National Accelerator Laboratory.*)

accelerators are the 200 to 500 GeV synchrotron at the National Accelerator Laboratory near Batavia, Illinois, and the 76-GeV synchrotron at the Institute for High Energy Physics near Serpukhov in the Soviet Union (Figs. 2-1 and 2-2). The highest energy electron accelerator is the 21-GeV linear accelerator at the Stanford Linear Accelerator Center (SLAC) (Fig. 2-3).

There are a variety of important reasons for building accelerators with the highest practical beam energies. Some of these are:

1. At high energies the shorter de Broglie wavelength h/p of the beam particles allows the possibility of probing the structure of target particles in increasingly fine detail.
2. Since there is as yet no general theory for the strong inter-

Figure 2-2 A small section of the 76-GeV accelerator at Serpukhov near Moscow. The vacuum pipe in which the beam travels is clearly visible on the right. A few of the many bending magnets can be seen to the left. (*Photograph courtesy of Professor Logunov, Institute for High Energy Physics, Serpukhov, USSR.*)

actions and only an incomplete one for the weak interactions, the high-energy behavior of cross sections, etc., can only be determined experimentally. Moreover there are reasons to hope that in the limit of very high energies, the strong interactions will be simpler to treat theoretically, which may allow an important breakthrough in their understanding. In any case it does not seem likely that the strong interactions will ever be understood completely until we can experimentally study their asymptotic behavior at very high energies.

3. The production of secondary particles (pions, kaons, antinucleons, neutrinos, etc.) is generally more copious for higher beam energies. This facilitates the study of their interactions and allows them to be studied at higher energies. It also becomes

Figure 2-3 Aerial view of the 21-GeV electron linear accelerator at SLAC. The overall length of the accelerator is 3 km (1.9 miles). The electrons begin their journey at the far end. The buildings in the foreground house the experimental areas. (*Photograph from SLAC.*)

Figure 2-4 A section of the CERN ISR. One of the beam intersection points is just above the center of the picture. Protons with energies up to 28 GeV are injected into the ISR from the CERN proton synchrotron. The proton beams can be stored in the ISR for many hours. (*Picture from PHOTO CERN.*)

practical to make beams of short-lived particles such as hyperons at high energies (Prob. 2-6), and thus make detailed studies of their interactions possible.

4. Cross sections for weak interactions generally increase with energy. This makes experiments to study neutrino interactions, for example, considerably easier at higher-energy accelerators.
5. There is of course the possibility that a completely new range of phenomena will be discovered at very high energies. This is the most exciting prospect, about which we can only speculate. Perhaps massive relatively stable particles will be found,[1] or a whole new class of interactions discovered.

In addition to the more conventional accelerators listed in Table 2-1, colliding-beam accelerators are becoming increasingly important. The CERN Intersecting Storage Rings (ISR) are shown in Fig. 2-4. In the ISR, two proton beams circulating in opposite directions are allowed to collide where the two rings intersect. Interaction rates are quite modest; however the ISR permits the study of proton-proton interactions at center-of-mass energies far greater than those attainable at any conventional proton accelerator.[2] Several electron-positron and electron-electron storage rings have also been built or are under construction. The most energetic of these are the storage rings at the Cambridge Electron Accelerator, at DESY in Hamburg, Germany, and at the Institute for Nuclear Physics in Novosibirsk, USSR with energies of about 3 GeV per beam.

2-2 THE PASSAGE OF CHARGED PARTICLES THROUGH MATTER

Here we shall briefly summarize the important aspects of the behavior of charged particles in matter to provide a basis for the material on detectors to be discussed later.

In addition to nuclear interactions, charged particles passing

[1] Possibilities here are the intermediate vector boson (Sec. 6-3) or the quark (Chap. 8).

[2] For a collision between a moving particle and one fixed in the laboratory, at very high energies the total energy in the c.m.s. is $\sim \sqrt{2ME}$, where E is the total laboratory energy of the incident particle, and M the mass of the target. For two protons colliding head on, each with total energy E in the laboratory, the total energy in the c.m.s. (which is at rest in the laboratory) is $2E$.

through matter lose energy through a variety of electromagnetic processes. For all charged particles except the electron, collisions with electrons are the most important process. (Electrons behave quite differently than the heavier particles because of their small mass and the absence of strong interactions. We shall discuss them separately below.)

ENERGY LOSS FOR HEAVY PARTICLES

In an encounter with an atomic electron a heavy, charged particle can excite it to a higher level or remove it completely from the atom. The trail of free electrons and ions thus formed by a charged particle forms the basis for most techniques for detecting such particles (e.g., bubble chambers, scintillation counters, spark chambers, photographic emulsions). Occasionally the electrons receive enough energy to make them visible as distinct tracks in bubble chambers and cloud chambers. These electrons are called *delta rays*.

The energy loss due to ionization can be calculated in a relatively straightforward fashion (see, for example, in Bibliography, E. G.

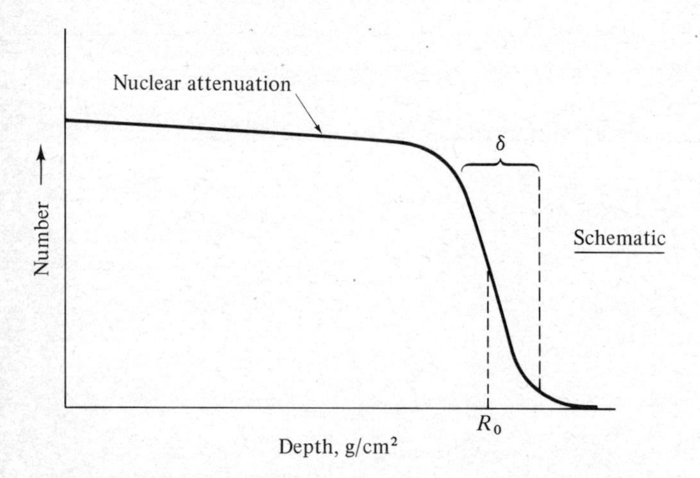

Figure 2-5 The variation of intensity of a beam of particles (initially monoenergetic) with depth. The particles gradually lose energy and eventually stop. For hadron beams there is also a continuous attrition of beam particles due to nuclear interactions. (The depth in grams per square centimeter is the product of the distance traversed and the density of the material.)

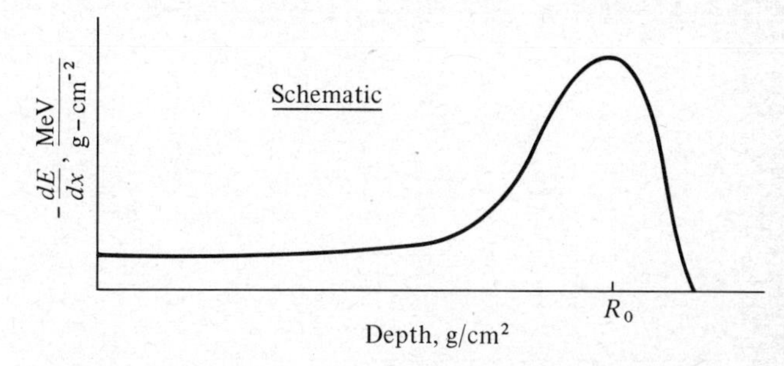

Figure 2-6 The variation in the rate of energy loss for a charged particle with depth. The rate of energy loss increases dramatically as the particle slows down near the end of its range.

Segrè, "Nuclei and Particles"). For our purpose a brief summary of the results will suffice. It is found that the rate of energy loss, dE/dx, of a particle is a function only of its velocity and its charge squared. The typical behavior of a beam of charged particles (initially monochromatic) in matter is shown in Fig. 2-5. For hadrons there is a continuous attenuation due to nuclear interactions in the medium; this does not occur with muon beams. On the average the particles have a rather well-defined range R_0, but statistical fluctuations in the energy loss cause some variation about R_0. This effect is called *range straggling;* δ is a measure of the fluctuation in range.

The behavior of dE/dx, the rate of energy loss, is shown in Fig. 2-6. For fast particles ($v \sim c$), the rate of loss is almost constant and rather small. As the particle slows down, however, $|dE/dx|$ rapidly increases. (There is also a slight rise in $|dE/dx|$ for ultrarelativistic velocities, but generally this is negligible.) For $v \approx c$ in hydrogen $-dE/dx \approx 4.1$ MeV/g-cm^{-2}, while for most other substances $-dE/dx \approx 2$ MeV/g-cm^{-2}. Tables of ranges and dE/dx for various materials are readily available.[1]

ENERGY LOSS FOR ELECTRONS

Because of their small mass, electrons lose a significant fraction of their energy by radiation or *bremsstrahlung* (braking radiation).

[1] See, for example, W. H. Barkas and M. J. Berger, *NASA Rep. SP* 3013 (1964), or B. Rossi, "High Energy Particles," Prentice-Hall, Englewood Cliffs, N.J., 1952.

Bremsstrahlung occurs when a charged particle is accelerated in the electric field of a nucleus. This effect is more important in high-Z materials. The average rate of energy loss due to radiation goes approximately as Z^2/m^2, where m is the rest mass of the incident particle, and Z the atomic number of the nucleus. Thus radiation loss is much less important for the heavier particles.

On the average a high-energy electron in traversing x g/cm² of material will lose a fraction of its energy

$$\frac{\Delta E}{E} \approx \frac{x}{X_0} \tag{2-1}$$

where X_0 is called the *radiation length* and is a function of Z. (For hydrogen, $X_0 = 58$ g/cm², while for lead, $X_0 = 6.5$ g/cm².) The actual energy loss suffered by a given electron varies widely about the mean, and therefore it is not possible to speak of a well-defined range for a high-energy electron in a high-Z material (for which radiation losses are much greater than collision losses).

MULTIPLE COULOMB SCATTERING

When a charged particle passes through matter, it undergoes a series of scatterings because of encounters with the electric fields of individual nuclei. These cause a deviation of the particle from its original direction. The deviation caused by a single encounter is likely to be small, but the cumulative effect of many encounters can be rather large. This has an important bearing on the precision attainable in measuring the angles or momenta of a particle in, say, a bubble chamber (Prob. 2-4). An approximate but useful formula for the rms angle of deviation is

$$\langle \theta^2 \rangle^{\frac{1}{2}} \approx \frac{21 \text{ MeV}}{p\beta c} \sqrt{\frac{x}{X_0}} \tag{2-2}$$

where θ is the space angle between the original direction of the particle and its direction after traversing a thickness x of the material, X_0 is the radiation length previously defined, $\beta = v/c$, and p is the momentum of the particle in units of MeV/c.

Pavel A. Čerenkov. Born 1904 in the Voronezh Region, USSR. He graduated from Voronezh State University in 1928. In 1934, while working under S. I. Vavilov at the Lebedev Institute of Physics, he observed the emission of blue light from a bottle of water when it was irradiated by a radium source. In a series of experiments Čerenkov showed that the light was not due to fluorescence as had been generally believed, but was a new type of radiation generated by fast, charged particles. In 1937 I. M. Frank and I. E. Tamm developed the theory of the Cerenkov effect. They showed that the radiation occurs when a charged particle moves through a transparent medium with a velocity greater than the velocity with which light propagates in the medium. In 1958 Čerenkov, Frank, and Tamm shared the Nobel Prize for their work. (*Photograph from the Niels Bohr Library, American Institute of Physics.*)

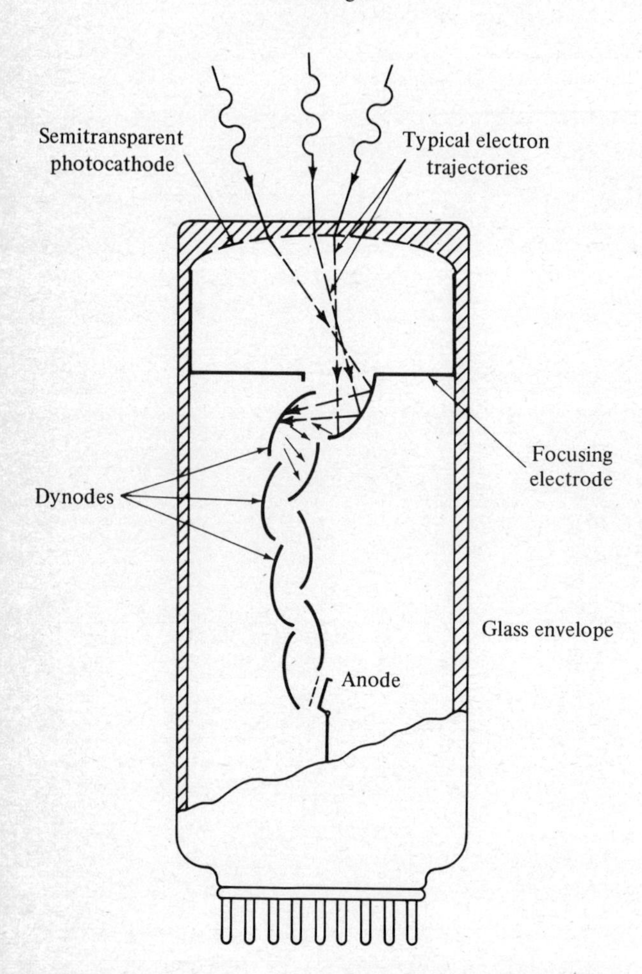

Figure 2-7 Schematic of a photomultiplier tube. Photomultipliers are invariably used to detect the light emitted from a scintillation or Cerenkov counter and convert it to an electrical pulse.

CERENKOV RADIATION

When a charged particle passes through a transparent medium with a velocity greater than that of light in the medium (that is, $\beta > 1/n$, where n is the index of refraction), it will lose a small fraction of its energy in the form of *Cerenkov radiation.* The process is analogous to the shock wave or sonic boom produced by a plane whose velocity is greater than that of sound in air. The Cerenkov light is emitted in a cone whose angle is $\cos \theta = 1/n\beta$ over a broad range of wavelengths. This light can be detected in suitable geometries with a photomultiplier tube (Fig. 2-7). Cerenkov detectors have proved very useful for measuring velocities of charged particles. They can work as threshold counters which respond only to particles with $\beta > 1/n$, or they can select a certain velocity band by making use of the directional properties of the light.

2-3 THE INTERACTION OF HIGH-ENERGY PHOTONS WITH MATTER

In passing through matter, high-energy photons (gamma rays) can lose energy by several means:

1. *Pair production.* The formation of an electron-positron pair in the field of a nucleus. This process is the predominant one for $E_\gamma \gg 1$ MeV.
2. *Compton scattering.* The elastic scattering of a photon by a free electron (or other charged particle).
3. *Photoelectric effect.* The process in which the photon gives up all its energy to a bound electron and removes it from the atom. Photoelectric absorption is typically strongly energy dependent, exhibiting a resonancelike behavior when the photon energy is just equal to that required to remove an electron from a given atomic shell.

For most elements photoelectric absorption is predominant for $E_\gamma \ll 1$ MeV. Compton scattering is important for $E_\gamma \sim 1$ MeV, and pair production predominates for $E_\gamma \gg 1$ MeV. As a rule of thumb the mean free path for a photon in a material is X_0, the radiation length previously defined. High-energy electrons or photons in matter typically initiate "showers" since the electrons produce photons by bremsstrahlung and the photons in turn produce pairs. This behavior provides a convenient means for distinguishing high-energy electrons from muons, which, because of their larger mass, do not radiate significantly.

2-4 THE SCINTILLATION COUNTER

The scintillation counter has become one of the most important devices for detecting fast particles. When a charged particle passes through certain materials, some of the energy lost in ionization goes into producing light in the visible part of the spectrum. This light can readily be detected by a photomultiplier tube. The most widely used scintillating material is polystyrene doped with certain organic scintillators. This material is relatively inexpensive and can be easily machined into almost any desired shape. Another important advantage of this type of scintillator is that the light is given off in

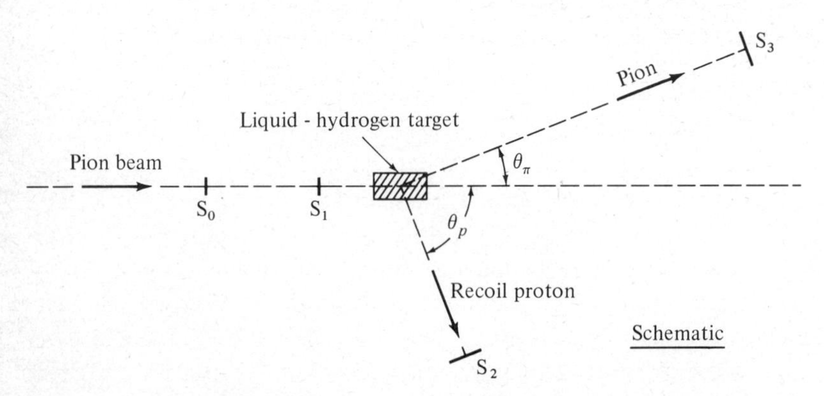

Figure 2-8 A possible arrangement of scintillation counters to study angular distributions in elastic π-p scattering. A pion beam is incident from the left. The observation of pulses from each of the counters S_0, S_1, S_2, and S_3 with the proper time relation indicates that an incident pion has entered the target and scattered at an angle θ_π, and a recoil proton has been detected at the appropriate angle θ_p.

Figure 2-9 A liquid-hydrogen target used in a scattering experiment. The vessel containing the liquid hydrogen is the smaller container barely visible between the "jaws" of the larger vacuum jacket. The bottom of the reservoir containing the liquid-hydrogen supply can be seen at the top of the picture. In this experiment the beam was incident from the left through the hole in the shielding wall. Thin mylar windows were used on the vacuum jacket to minimize the material the scattered protons passed through. The mylar windows are wrinkled because of the vacuum loading.

approximately 10^{-9} sec. This makes it possible to determine particle velocities by measuring the transit time between pairs of counters spaced some distance apart. Through the use of two or more counters which are electronically required to give pulses at the proper times, it is possible to select certain kinds of events even in the presence of a large background of undesirable radiation. As a simple example, one can study angular distributions in π-p scattering by using an arrangement like that shown in Fig. 2-8. A fast coincidence between pulses from scintillation counters S_0, S_1, S_2, S_3 (with appropriate delays in the signals to compensate for transit times) indicates that a beam particle of approximately the right velocity entered the target and scattered at an angle θ_π, and a recoil proton came off at the appropriate angle θ_p. A hydrogen target suitable for use in such an experiment is shown in Fig. 2-9. By suitable arrangements of counters and bending magnets it is possible to measure momenta.

An important use of scintillation counters is to trigger spark chambers (described below). Here the counters are used to crudely constrain the geometry of the event, and the spark chambers provide much better spatial resolution. Thus an experimenter has a means of photographing or otherwise recording only those events which have a configuration like the desired one, and uninteresting events can be disregarded.

2-5 SPARK CHAMBERS

When a charged particle passes through a gas, it leaves behind a trail of ion pairs. This trail can be made visible by applying an electric field with a suitable arrangement of electrodes immediately after the particle passes. Under proper conditions a visible electrical discharge occurs along the track.

A typical spark chamber (Fig. 2-10) consists of a series of parallel metal foils. Alternate foils are grounded. The other plates can be brought quickly to high voltage by a suitable switching arrangement (usually a spark gap or thyratron). The gas is usually a helium-neon mixture which has been found to give good tracks if the high-voltage pulse is applied within approximately 1 μsec after the particle passes. The chamber(s) are generally triggered by an arrangement of scintillation counters which serve to impose the desired constraints on the particle trajectories. The sparks can be recorded photographically, and precise measurements of trajectories made from the film. Figure 2-11 shows an actual spark-chamber array in operation.

Spark chambers with plates of brass, lead, or other heavy materials can be used to detect gammas, neutrons, and neutrinos which interact in the plates to produce charged particles.

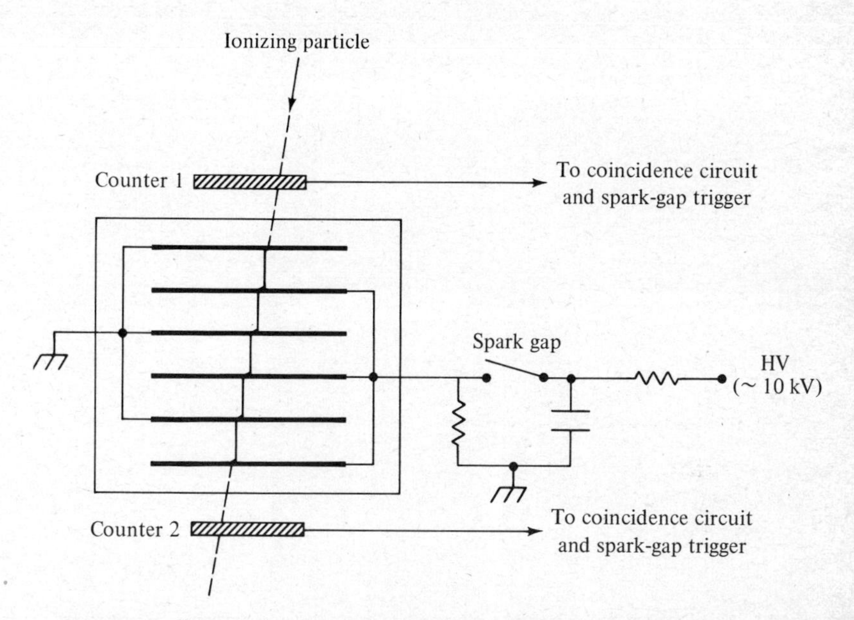

Figure 2-10 Schematic of a spark chamber. When a coincidence between the two scintillation counters is observed, the spark gap is triggered, which applies high voltage to alternate plates of the chamber. With proper operating conditions sparks will occur along the path of the particle.

2-6 BUBBLE CHAMBERS

D. A. Glaser discovered that when charged particles pass through a superheated liquid, bubbles tend to form along their tracks.[1] If the liquid is photographed after an appropriate delay, the paths of the particles and their interactions in the liquid show up very clearly as tracks composed of tiny bubbles.

The liquid most used in bubble chambers is liquid hydrogen. This has the great advantage that all target particles are free protons. Sometimes neon is mixed with the hydrogen to reduce the radiation length and thereby increase the probability of seeing γ's (mostly from π^0 decays) in the chamber by causing them to produce electron-positron pairs. Other useful operating liquids are propane and propane-freon mixtures. These have the advantage of higher density and much higher probability of γ detection. Disadvantages are

[1] D. A. Glaser, *Phys. Rev.* **87,** 665 (1952); *Sci. Am.*, February, 1955, p. 46.

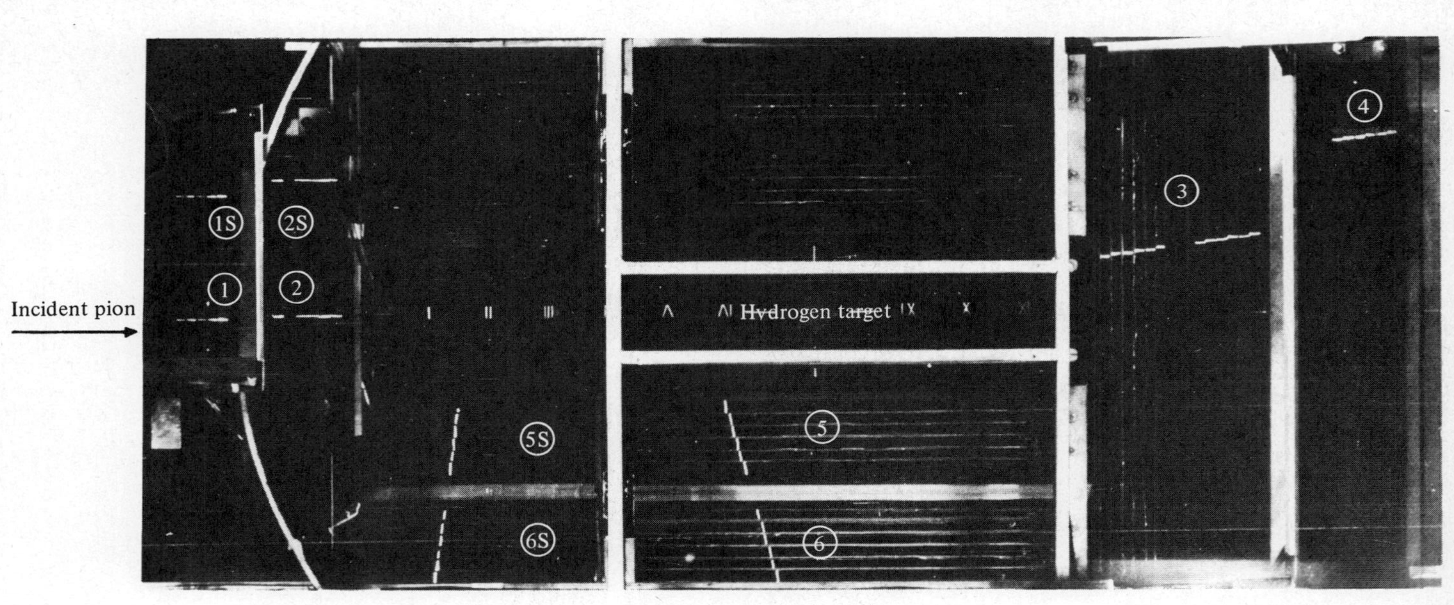

Figure 2-11 An actual spark-chamber photograph showing the elastic scattering of a 3-GeV/c π^+ from a proton. The pion is incident from the left, and its path is recorded in chambers 1 and 2. The pion scatters off a proton in the 45-cm-long hydrogen target. The scattered pion is recorded in chambers 3 and 4. (Chamber 4 is considerably farther from the target than it appears here; the views of the various chambers are brought together by a system of mirrors.) The recoil proton appears in chambers 5 and 6. Views 1S, 2S, 5S, and 6S are the 90° stereo views of chambers 1, 2, 5, and 6 respectively. (Compare Fig. 2-8.) (*Photograph courtesy of Professor L. W. Jones, University of Michigan.*)

Donald A. Glaser. Born 1926 in Cleveland, Ohio. Glaser received his Ph.D. from the California Institute of Technology in 1949 and that year joined the faculty at the University of Michigan. There he began to experiment with an idea to use a superheated liquid to detect high-energy particles. He first showed that superheated diethyl ether erupted into violent boiling when exposed to a radioactive source. Before long he was able to observe individual tracks in the liquid. The eventual development of large liquid-hydrogen bubble chambers provided physicists with a revolutionary new tool for studying elementary particles and their interactions. Glaser received the Nobel Prize in 1960 for his work. Shortly thereafter he became interested in molecular biology and is now active in that field. (*Photograph from the Niels Bohr Library, American Institute of Physics.*)

Luis W. Alvarez. Born 1911 in San Francisco. He received his doctorate in 1936 from the University of Chicago. During the war he worked on the Manhattan Project and radar research. Alvarez was codiscoverer of H^3 and He^3 and made important contributions to the development of proton linear accelerators. Upon learning of Glaser's pioneering work with bubble chambers the increased coulomb scattering of the particles, which makes measurements of angles and momenta considerably less accurate, and the fact that the target particles are bound in a nucleus. Liquid deuterium is often used instead of hydrogen when interactions with neutrons are to be studied.

Bubble chambers have grown enormously since Glaser's first chambers, whose dimensions were on the order of several centimeters. The largest bubble chambers now in operation have dimensions of several meters and useful volumes of about 20 m^3 (Fig. 2-12). A large magnetic field ($\sim$20 kG) is applied to allow the momenta to be determined from the curvatures of the tracks. Most modern chambers use superconducting coils to produce the magnetic field.

Unlike the operation of a spark chamber, the expansion of a bubble chamber cannot be triggered by counters. The liquid is normally below the boiling point. It is rapidly raised above the boiling point by reducing the pressure just before the burst of particles enters. The lights are flashed and a picture taken just after the particles pass through. The liquid is then recompressed, and the chamber is ready for another cycle.

As a tool for studying particle interactions the bubble chamber has the advantages of showing the vertex (interaction point) and all charged particles. Furthermore it possesses inherently good spatial resolution and nearly isotropic response. Its main disadvantage, compared with a spark chamber, is its lack of selectivity. One just fires particles in and takes pictures. This makes the bubble chamber

---

he immediately recognized their potential usefulness in elementary particle research and began a program to develop large liquid-hydrogen chambers. He and his group at the Lawrence Radiation Laboratory (now the Lawrence Berkeley Laboratory) built a succession of larger and larger chambers, culminating in 1959 with the successful operation of a 72-in. chamber. For many years this was the largest chamber in the world. The same group also pioneered in the development of scanning and measuring machines and of computer programs for rapid analysis of the millions of pictures which eventually poured from these chambers. In 1968 Alvarez was awarded the Nobel Prize for his development of hydrogen bubble chambers and data analysis techniques and for his participation in the discovery of a large number of short-lived particles, resonances. In the photograph Alvarez is holding the first liquid-hydrogen chamber built under his direction. The vacuum chamber for a larger, 15-in. chamber is at the right. The glass window for the 72-in. chamber is in the foreground. (*Photograph from the Lawrence Berkeley Laboratory.*)

Figure 2-12 Assembly of the world's largest heavy-liquid bubble chamber *Gargamelle* at CERN. The chamber, with a useful volume of 10 m^3, will be used principally for neutrino experiments. Here the chamber body is being installed inside the magnet coils. (*From PHOTO CERN.*)

most useful in a survey type of experiment where the grosser features of, say, π^+-p interactions are studied at various energies. However, spark chambers or scintillation counters are generally more useful for studying a certain kind of event (say, π^+-p elastic scattering) in great detail or for studying interactions with small cross sections, especially when intense beams of the particles to be studied are available (for example, e-p elastic scattering at large angles). Also spark-chamber and counter techniques are generally more flexible and lend themselves well to special situations such as the study of polarization by using polarized targets or double scattering (Sec. 3-3).

Another important advantage of spark chambers and especially of counters is that the analysis of the data is considerably easier. The analysis of a half-million bubble-chamber pictures is a formidable task, generally needing some man-years of effort by scanners, measurers, and physicists, and many hours of computer time for computations. Generally spark-chamber pictures are considerably simpler and faster to analyze. Also systems for reading out spark-chamber coordinates electronically have been developed to simplify the task further.

An important disadvantage of spark-chamber–counter techniques is that it is generally not possible to see the vertex. It is also very difficult to build a detector with equal sensitivity in all directions.

2-7 OTHER TECHNIQUES

At one time, cloud chambers and photographic emulsions were important tools in the study of particle interactions. They have largely been superseded by bubble and spark chambers.

PROBLEMS

See Appendix B for a summary of results from relativistic kinematics and tables of kinematics.

2-1 Calculate the threshold energy for the reaction $p + p \rightarrow p + p + \bar{p} + p$ if the target proton is at rest.

2-2 (*a*) Calculate the total energy in the c.m.s. for a collision of a 20-GeV proton with one at rest. Compare this with the head-on collision of two 20-GeV protons.[1]

(*b*) Do the same for 20-GeV electrons.

2-3 A relativistic particle loses approximately 1.7 MeV/g-cm^{-2} in neon. If it takes, on the average, 37 eV to produce an ion pair, estimate how many ion pairs are produced per centimeter in neon gas at atmospheric pressure.

2-4 Calculate the rms scattering angle due to multiple coulomb scattering for a 2-GeV pion traversing 10 cm of liquid propane ($\rho = 0.41$ g/cm^3; $X_0 = 44.6$ g/cm^2). Compare this with the angle through which the pion is bent if it travels 10 cm in a magnetic field of 20 kG. Discuss the implications of this comparison for the momentum resolution in a propane bubble chamber operating under these conditions. Repeat for liquid hydrogen ($\rho = 0.06$ g/cm^3; $X_0 = 58$ g/cm^2).

2-5 What would be a suitable radiator for a threshold Cerenkov counter that is to be sensitive to 800 MeV/c pions but not to 800 MeV/c protons? To 8 GeV/c pions but not to 8 GeV/c protons?

2-6 High-energy experiments typically have dimensions $\sim$1 m. Discuss the feasibility of forming beams of μ^+, π^+, K^+, Λ^0, and π^0 and transporting them over distances of about 1 m. In particular, calculate the beam momentum for which the mean free path for decay is 1 m for each of the above particles.

[1] Unless specified otherwise, all energies given in the text are kinetic, rather than total, energies.

CHAPTER 3

discussion of strong interactions

There are therefore Agents in Nature able to make the Particles of Bodies stick together by very strong Attractions. And it is the Business of experimental Philosophy to find them out.

From Sir Isaac Newton,
"Opticks," Dover, New York, 1952

3-1 INTRODUCTION

The strong interaction is most familiar as the force between nucleons responsible for nuclear binding. Generally speaking, mesons, nucleons, and hyperons interact among themselves through strong interactions. Collectively, as stated in Chap. 1, such particles are called hadrons. As a rule, if a reaction satisfies the conservation laws appropriate to the strong interactions, it is likely to occur with appreciable probability. For example, at very high energies, rates for the reactions

$$
\begin{aligned}
&p + p \rightarrow n + p + \pi^+ \\
&p + p \rightarrow p + p + \pi^0 \\
&p + p \rightarrow p + p + \pi^+ + \pi^- \\
&\cdots
\end{aligned}
$$

are of the same order of magnitude. Final states containing strange particles are somewhat less likely, but still appear with appreciable probability.

Unfortunately no satisfactory theory of strong interactions has yet been advanced. There is nothing equivalent to the combination of quantum mechanics and electromagnetic theory (*quantum electrodynamics*), which has proved so successful in atomic physics and, in principle at least, seems capable of "explaining" all atomic physics. The difficulty in the theory of the strong interactions seems basically due to the strength of the interaction itself. As a result approximations, such as perturbation theory in quantum mechanics, break down. Furthermore, because of a phenomenon known as *virtual dissociation,* all the hadrons are strongly coupled to each other. Thus for a time $\Delta t \sim \hbar/\Delta E$ (where ΔE is the apparent energy viola-

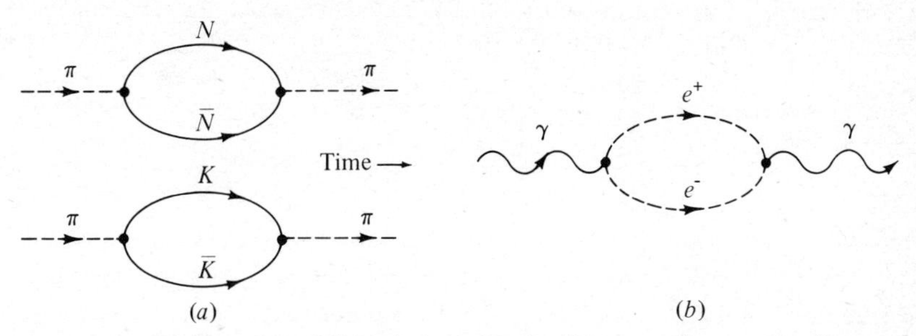

Figure 3-1 (a) The virtual dissociation of a pion into a nucleon-antinucleon or K-$\bar{K}$ pair. (b) The dissociation of a photon into an e^+-e^- pair.

tion) allowed by the uncertainty principle, the π meson can be thought to exist as a nucleon-antinucleon pair or a K-$\bar{K}$ pair (Fig. 3-1a). The photon can likewise dissociate into an electron-positron pair through virtual electromagnetic interactions (Fig. 3-1b), but the probability of this dissociation is much lower than for those occurring through virtual strong interactions, because electromagnetic interactions are inherently weaker. Thus virtual electromagnetic processes can often be neglected, and even when it is necessary to take them into account, there exist suitable calculational techniques.

Because of such virtual processes the hadrons are all closely coupled to each other and, in a sense, every problem in strong interactions is a many-body problem. Feynman has described the nuclear force as "almost as complicated as it can be." It depends on velocity and spin orientation as well as position and isotopic spin. Virtual strong interactions are also a problem in treating the weak decays of hadrons since, for example, a K meson can be thought of as a virtual $\bar{\Lambda}^0$-nucleon pair. This has hampered the development of a complete theory of the weak interactions, which otherwise would be amenable to approximation schemes of the type used for electromagnetic interactions.

The bulk of the experimental data on strong interactions has been obtained by studying the scattering of one particle on another, for example,

$$p + p \to ?$$
$$\pi + p \to ?$$

It is possible to make fairly intense beams of $\pi^\pm$, K^0, $K^\pm$, n, p, and $\bar{p}$ for such studies. The most accessible target particles are protons (in the form of a liquid-hydrogen target). Scattering from neutrons can be studied somewhat less directly through

$$p + d \rightarrow ?$$
$$\pi + d \rightarrow ?$$

Since the deuteron is a loosely bound n-p system, in many situations the neutron can be considered a free neutron with corrections made to account for the effects of the proton nearby.

In general it is not possible to study systems like the π-π system directly in scattering experiments, but some information can be gleaned indirectly (Sec. 3-7).

3-2 DEFINITION OF A CROSS SECTION

The probability of a certain reaction can be expressed in terms of a *cross section.* If we imagine a beam of particles passing through a thickness dx of a target, the fraction of particles that interact because of collisions can be written

$$\frac{dI}{I} = -\sigma N\,dx \tag{3-1}$$

where I is the beam flux in appropriate units (e.g., total particles or particles per second), and N is the number of target nuclei per cubic centimeter in the target. The cross section σ is a proportionality constant with dimensions of area. Common units are

$$1 \text{ barn (b)} = 10^{-24}\text{ cm}^2 \qquad 1\text{ mb} = 10^{-27}\text{ cm}^2$$

The cross section σ can be thought of as the effective area of a nucleus (Fig. 3-2). This might be very different from the geometric area. We can define various types of cross sections. For example,

Total cross section σ_T. The cross section for all processes which scatter or otherwise remove particles from the beam

Elastic cross section σ_{el}. The cross section for elastic scattering, for example, $p + p \to p + p$

Inelastic cross section σ_{inel}. The cross section for all processes other than elastic scattering, $\sigma_{\text{inel}} = \sigma_T - \sigma_{\text{el}}$

We can also define a differential cross section $d\sigma/d\Omega$ by the equation[1]

$$\frac{\Delta I}{I} = \left(\frac{d\sigma}{d\Omega}\Delta\Omega\right) Nx \qquad (3\text{-}2)$$

Here x is the target thickness, and $\Delta I/I$ is the fraction of the incident beam scattered into a certain solid angle $\Delta\Omega$ (Fig. 3-3). Generally $d\sigma/d\Omega$ will be a function of angle and incident energy. Unless otherwise specified $d\sigma/d\Omega$ will refer to the cross section in the c.m.s. It is often convenient to define the differential cross section in such a way that it is relativistically invariant. This can be done by defining the differential cross section in terms of the invariant four-momentum transfer t, where t is the square of the difference between the final and

[1] We can define a differential cross section for elastic or inelastic scattering, but unless otherwise specified we refer to the differential cross section for *elastic* scattering.

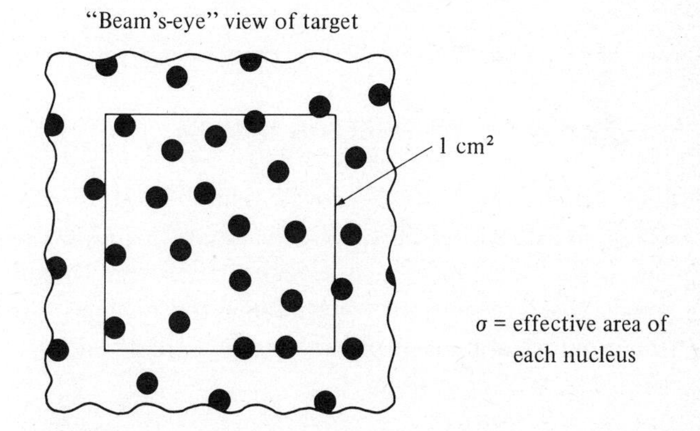

Figure 3-2 A "beam's-eye" view of a target. The shaded circles represent the effective area of each nucleus. If a beam particle strikes the shaded area, it is assumed to have interacted. The fraction of the beam that interacts, dI/I, is therefore the ratio of the shaded area to the total area. In the 1-cm² area there are $N\,dx$ targets, each of area σ, so that $-dI/I = \sigma N\,dx/(1\ \text{cm}^2)$.

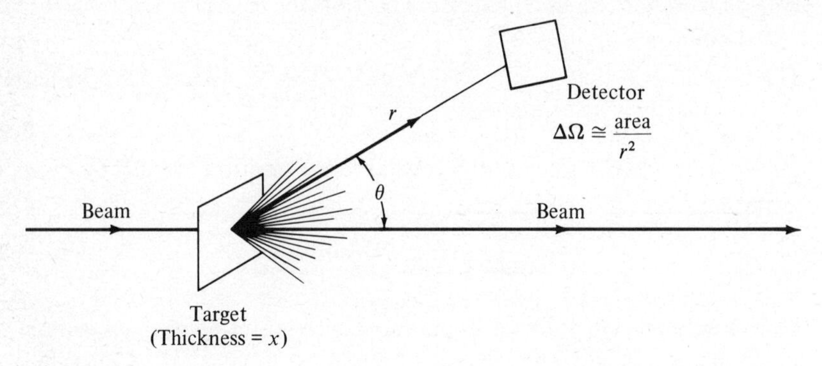

Figure 3-3 Basic arrangement for a differential cross-section measurement. The beam is incident on a thin target. The detector, which subtends a solid angle $\Delta\Omega$ (assumed small), detects particles coming off at an angle θ.

initial four-momentum of the incident particle.[1] For elastic scattering, $t = -2p^{*2}(1 - \cos\theta_{\text{c.m.}})$, where p^* is the momentum of either particle in the c.m.s., and $\theta_{\text{c.m.}}$ the c.m.s. angle. The differential cross section $d\sigma/d|t|$ is then related to the cross section in the c.m.s., $d\sigma/d\Omega$, by

$$\frac{d\sigma}{d|t|} = \frac{\pi}{p^{*2}} \frac{d\sigma}{d\Omega} \tag{3-3}$$

3-3 THE NUCLEON–NUCLEON SYSTEM

The nucleon-nucleon (N-N) system has been studied very extensively. Its great importance lies in the fact that the properties of this system determine those of the nucleus. Much can therefore be learned about the N-N system by studying the nucleus. However this is the province of nuclear physics, and we shall restrict ourselves to the elementary N-N system.

[1] See Appendix B for a summary of relativistic formulas. Often q^2 is used for the invariant four-momentum squared, instead of t. There is also some inconsistency in the sign conventions used. As usually defined, t is negative for elastic scattering. In Sec. 5-1 we use q^2 for the absolute value of t to conform with common usage in electron-scattering experiments. An expression for q^2 is developed there in terms of the kinetic energy of the target particle after the scattering.

REVIEW OF EXPERIMENTAL DATA

We first review some of the basic experimental data. Figure 3-4 shows the behavior of the total cross section σ_T, the elastic cross section σ_{el}, and the inelastic cross section σ_{inel} for *p-p* scattering. Below the threshold for single-pion production $\sigma_{el} = \sigma_T$. [Note that a geometric cross section of 50 mb corresponds to a radius of 1.25×10^{-13} cm $\equiv$ 1.25 fermi (fm).[1]]

Cross sections for *n-p* scattering are shown in Fig. 3-5. The inelastic and elastic cross sections for *n-p* scattering are rather poorly

[1] A convenient unit of length in elementary particle physics is the *fermi*, where 1 fm $\equiv 10^{-13}$ cm. This corresponds roughly to the charge radius of the proton ($\approx$0.7 fm) or the Compton wavelength of the pion ($\approx$1.4 fm).

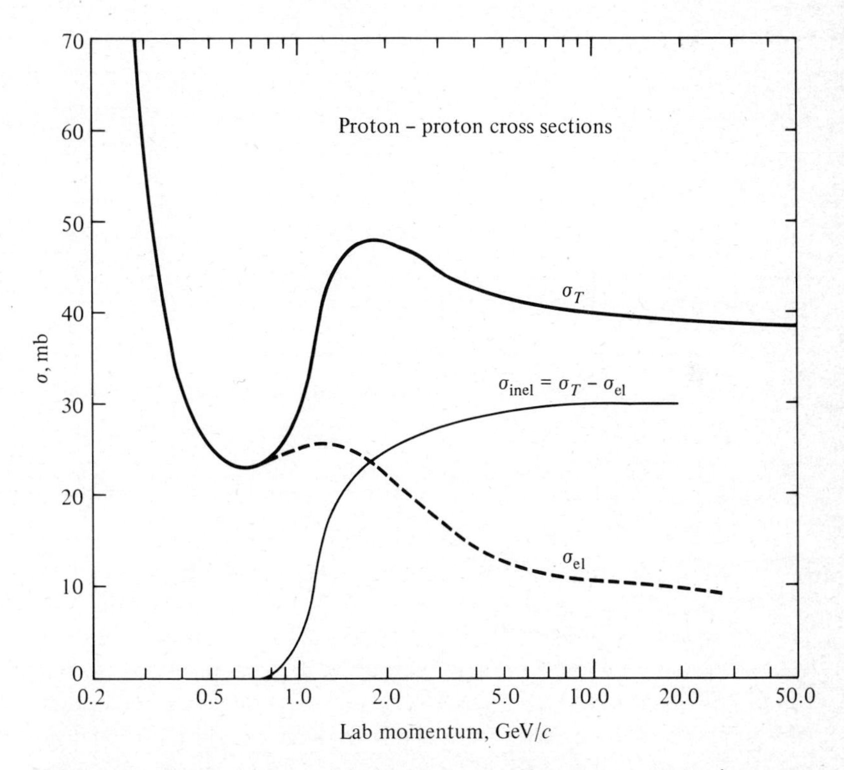

Figure 3-4 Total, elastic, and inelastic cross sections for *p-p* scattering versus momentum. (*Data taken from a compilation by the Particle Data Group, Lawrence Radiation Laboratory Rep. UCRL*-20,000 *NN*, *August*, 1970.)

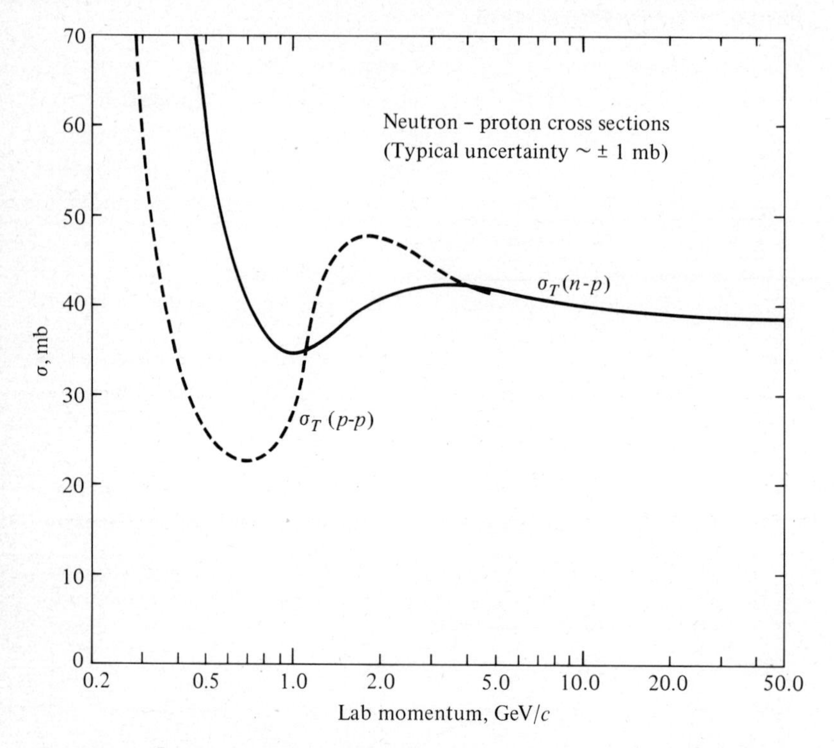

Figure 3-5 Neutron-proton total cross section versus momentum. The dashed line shows the *p-p* total cross section for comparison.

known and are not given. A common feature of total cross sections for all interactions is that at very high energies they become almost energy independent.

Another basic piece of experimental information is the differential cross section $d\sigma/d\Omega$ for elastic scattering.

If we consider a *p-p* collision in the c.m.s., it is clear from the symmetry of the initial state that all experimental quantities (for example, $d\sigma/d\Omega$) must be symmetrical about 90° in the c.m.s. Below approximately 500 MeV (kinetic energy in the laboratory system) the *p-p* differential cross section is almost independent of angle, but rises at small angles as a result of the long-range coulomb interaction between protons (Fig. 3-6).

At higher energies, as inelastic scattering becomes important, the differential cross section becomes more and more peaked near the forward direction (Fig. 3-7). This forward peak is often called the *diffraction peak.* It is basically a wave phenomenon, and is completely analogous to the diffraction of light around a spherical obstacle. The angular width of the diffraction peak is $\sim\lambda/R$, where R is the radius of the obstacle (just as it is in the more familiar case of a circular aperture such as a telescope).

The differential cross section for *n-p* scattering (Fig. 3-8) is generally similar to that for *p-p*. However since there is no longer symmetry about 90° in the c.m.s., we need to plot the cross section from 0 to 180°. At low energies the cross section is roughly symmetrical about 90°. At higher energies the diffraction peak becomes prominent at small angles. At high energies there is also a small,

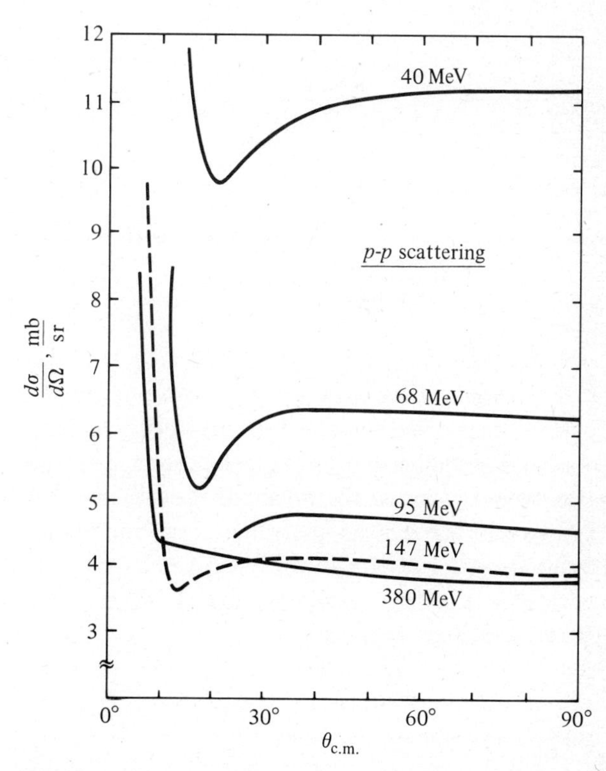

Figure 3-6 Typical behavior of the differential cross section for elastic *p-p* scattering as a function of c.m.s. angle at low energies.

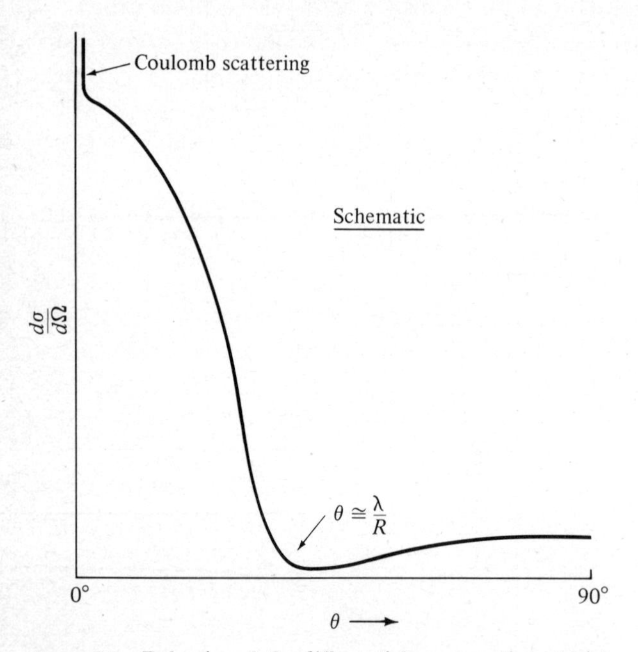

Figure 3-7 Behavior of the differential cross section at high energies (schematic). Most of the scattering is at angles $\lesssim \lambda/R$, where $\lambda = h/p$, and R is the radius of the target ($\sim$1 fm for protons).

very sharp peak in the cross section near 180° (see the 1-GeV curve in Fig. 3-8). This peak is considerably narrower than the forward peak. In the n-p system, 180° scattering in the c.m.s. corresponds to a head-on collision; the target particle after the collision has the same velocity and angle the incident particle had before the collision. It is just as though the neutron and proton had exchanged identities by the transfer of one unit of positive charge from the proton to the neutron. For this reason n-p scattering near 180° is referred to as *charge-exchange* scattering.

APPLICATION OF ISOTOPIC SPIN

A variety of evidence from nuclear physics suggests that the forces between nucleons are charge independent if allowances are made for

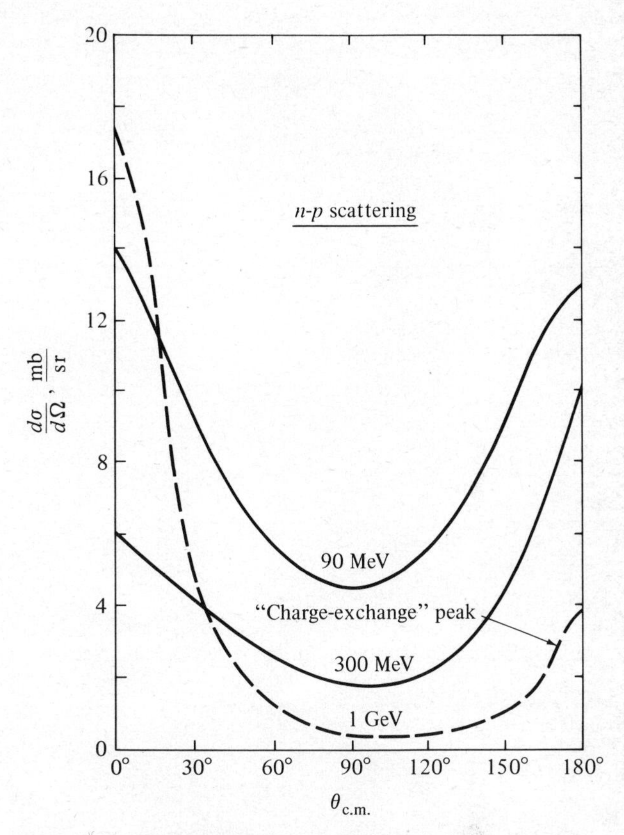

Figure 3-8 Typical behavior of the differential cross section for elastic n-p scattering.

coulomb effects (see, for example, E. G. Segrè, "Nuclei and Particles," sec. 10-4). This concept has been generalized in terms of isotopic spin to include similar regularities observed in other systems, such as the π-N system. As discussed in Sec. 1-8, the strong interactions depend only on the magnitude of the total isospin **T** of a system and not on its orientation in isospin space. In particular they do not depend on T_3, which is related to the charge by Eq. (1-1),

$$T_3 = \frac{Q}{e} - \frac{B}{2} - \frac{S}{2}$$

Emilio Segrè. Born 1905 in Tivoli, Italy. He received his doctorate in 1928 at the University of Rome under Enrico Fermi. He worked with Fermi on experiments to study artificial radioactivity induced by neutron bombardment. Segrè participated in the discovery of several elements, technetium, astatine, and plutonium 239. In 1938 he left Italy and began work at the University of California in Berkeley. He and his associates carried out a series of experiments to study nucleon-nucleon scattering. In 1959 Segrè and Chamberlain were awarded the Nobel Prize for their discovery of the antiproton in an experiment carried out with the Bevatron. (*Photograph from Meggers Gallery of Nobel Laureates, American Institute of Physics.*)

Owen Chamberlain. Born 1920 in San Francisco. During World War II Chamberlain was in the Manhattan Project. He received his doctorate in 1949 from the University of Chicago, where he worked under Enrico Fermi. Later he went to the University of California and worked with E. Segrè on nucleon-nucleon scattering experiments. He and Segrè shared the Nobel Prize for their discovery of the antiproton. Much of Chamberlain's efforts in recent years have been devoted to studies of polarization in nucleon-nucleon and pion-nucleon scattering. (*Photograph from the Niels Bohr Library, American Institute of Physics.*)

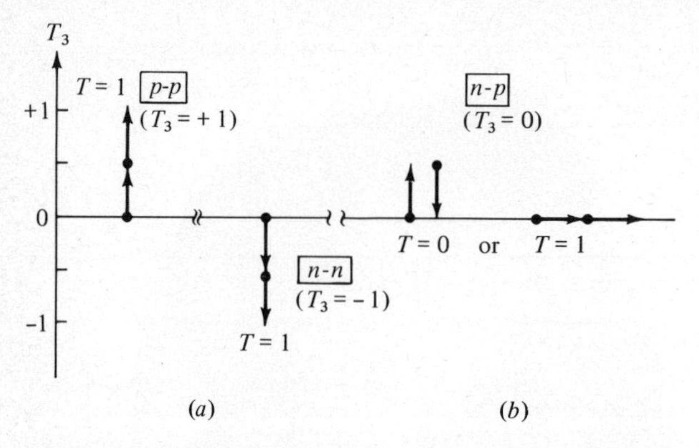

Figure 3-9 (*a*) Possible orientations of the isotopic spin vectors in isospin space for *p-p* and *n-n* systems. Since $|T_3| = 1$, the vector **T** must be one unit long. (*b*) Possible orientations of the isotopic spin vectors for the *n-p* system. Since $T_3 = 0$, the vector **T** can be either zero or one unit long.

This relation holds for a single particle or a system of particles. The proton has $T_3 = \frac{1}{2}$ (Table 1-2), and the neutron $T_3 = -\frac{1}{2}$. The *p-p* system has $T_3 = +1$, and the *n-n* system $T_3 = -1$. For both systems, $T \equiv |\mathbf{T}| = 1$ is the only possibility, so they are equivalent in their strong interactions (Fig. 3-9*a*).

The *n-p* system is somewhat more complicated. From the above relation for T_3, we find $T_3 = 0$ for the *n-p* system. However this can come about either because the isospins of the individual nucleons are antiparallel (so that $T = 0$), or because the vector **T** is at right angles to the T_3 axis (Fig. 3-9*b*). In other words, the general two-nucleon system can have $T = 0$ corresponding to isospins antiparallel, or $T = 1$ corresponding to isospins parallel. For $T = 1$, there are three possible orientations of **T**: parallel to the T_3 axis, antiparallel, or at right angles. (It is instructive to consider the possible orientations in combining ordinary spins in the two-electron system, a completely analogous situation.) Since the value of T, and not T_3, determines the strong interactions of a system, we can consider the *n-p* system as a linear combination of $T = 0$ and $T = 1$ states. We can express the wave function or state vector for a nucleon-nucleon system $|N,N\rangle$ in terms of eigenstates of isotopic spin

$|T,T_3\rangle$ as follows:[1]

$$\begin{aligned}|p,p\rangle &= |1,1\rangle \\ |n,n\rangle &= |1,-1\rangle \\ |n,p\rangle &= \frac{1}{\sqrt{2}}(|1,0\rangle - |0,0\rangle)\end{aligned}$$

In terms of the isotopic spin eigenstates, the invariance of strong interactions with respect to isotopic spin means that as far as strong interactions are concerned, the eigenstates $|1,1\rangle$, $|1,-1\rangle$, and $|1,0\rangle$ are equivalent.

As an example of the application of isospin invariance, consider the reaction

$$p + p \rightarrow d + \pi^+$$

The left-hand side has isospin 1. Empirically the deuteron has isospin 0, and the π has isospin 1, so that the right-hand side also has isospin 1. If we compare this with

$$n + p \rightarrow d + \pi^0$$

we note that the right-hand side is a pure $T = 1$ state, while the left-hand side is now an equal mixture of $T = 0$ and $T = 1$. But since the $T = 0$ part of the mixture cannot participate in the reaction (strong interactions conserve T), we conclude that

$$\sigma(n + p \rightarrow d + \pi^0) = \tfrac{1}{2}\sigma(p + p \rightarrow d + \pi^+)$$

Furthermore, since the differential cross section for $p + p \rightarrow \pi^+ + d$ must be symmetrical about 90° in the c.m.s., that for $n + p \rightarrow d + \pi^0$ must also be, since the above relation is true at every angle.

We shall consider some additional applications of isospin when we discuss the π-N system, for which isospin has more important implications.

[1] We use the standard bracket notation for representing a wave function; any labels used to specify the state are enclosed in the brackets.

EFFECTS OF SPIN

Nucleon spin considerably complicates the N-N interaction, both theoretically and experimentally. The N-N interaction, in general, depends on the relative orientation of the spins of the nucleons. Spin effects can be studied in scattering experiments with polarized beams or polarized targets, or both. The simplest experiment to study spin effects in the N-N interaction can be done in any of three essentially equivalent ways:

1. By scattering an unpolarized beam from a polarized target
2. By studying the polarization attained by an unpolarized beam after scattering from an unpolarized target
3. By scattering a polarized beam from an unpolarized target

Polarized proton targets have been available only in the past few years. Before that, most high-energy polarization experiments were done with a combination of methods 2 and 3 above. An unpolarized proton beam was scattered off a target, and the polarization of the scattered protons was then measured by scattering them off a second target (Fig. 3-10). After scattering from the first target at an angle θ_1, the protons will have a polarization

$$P_1(\theta_1) \equiv \frac{N_{\text{up}} - N_{\text{dn}}}{N_{\text{up}} + N_{\text{dn}}} \tag{3-4}$$

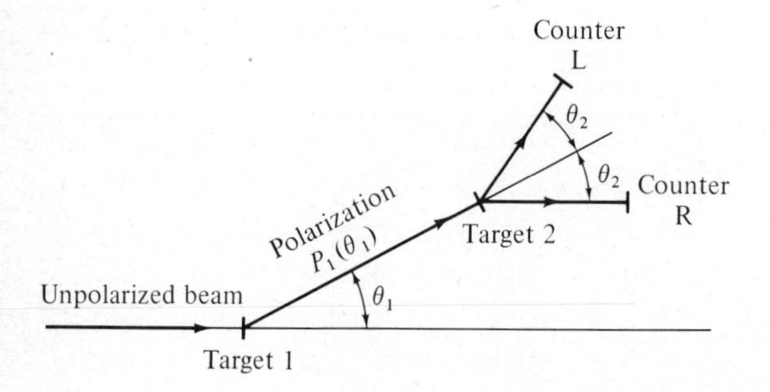

Figure 3-10 Schematic of a double-scattering experiment. An initially unpolarized beam scatters at an angle θ_1. After the first scattering, the beam has a polarization $P_1(\theta_1)$. Upon scattering in the second target, there will generally be a left-right asymmetry.

where N_{up} is the number of scattered protons with spin up (relative to the scattering plane), and N_{dn} the number with spin down. If the first target is a hydrogen target, $P_1(\theta_1)$ is determined by the spin dependence of the p-p interaction. The polarized proton beam incident on the second target will in general exhibit a preference for scattering either to the left or to the right (i.e., into counter L or counter R in Fig. 3-10). The left-right asymmetry $(L - R)/(L + R)$ will be determined by the polarization of the incident protons and the spin dependence of the second scattering. It is possible to show from time-reversal invariance that the polarization produced when an unpolarized beam scatters through a given angle is equal to the asymmetry observed when a completely polarized beam scatters through the same angle. Then if $\theta_1 = \theta_2$ and the energy loss in the first scattering can be neglected,[1] the asymmetry ϵ is

$$\begin{aligned}\epsilon &\equiv \frac{L - R}{L + R} \\ &= [P_1(\theta_1)]^2\end{aligned}$$

where L represents the number of scattering events detected in counter L, and R those detected in R. Usually one of the targets is not a hydrogen target, and it is then necessary to determine the beam polarization (or the asymmetry produced with a given beam polarization) in a separate experiment.

Polarization experiments are generally easier with polarized proton targets, though the polarized targets themselves are major pieces of apparatus. Unfortunately it has not been possible to polarize pure hydrogen. Presently operating targets use various organic compounds, such as ethylene glycol and butanol, or hydrated lanthanum magnesium nitrate. In the latter the protons in the water of hydration are polarized. At best, free protons constitute about 13 percent of the sample by weight. The target crystals are placed in a magnetic field, typically 18 to 40 kG, and cooled to temperatures $\lesssim 1°$K. The proton polarization at thermal equilibrium under these conditions is quite small, $\approx e^{-\mu_p B/kT}$, where μ_p is the proton magnetic moment, and B the magnetic field. (For $T = 1°$K and $B = 20$ kG, this works out to about 0.2 percent polarization.) However under

[1] This also assumes that both targets are hydrogen targets and that the detectors L and R are small so that both polar angles are well defined.

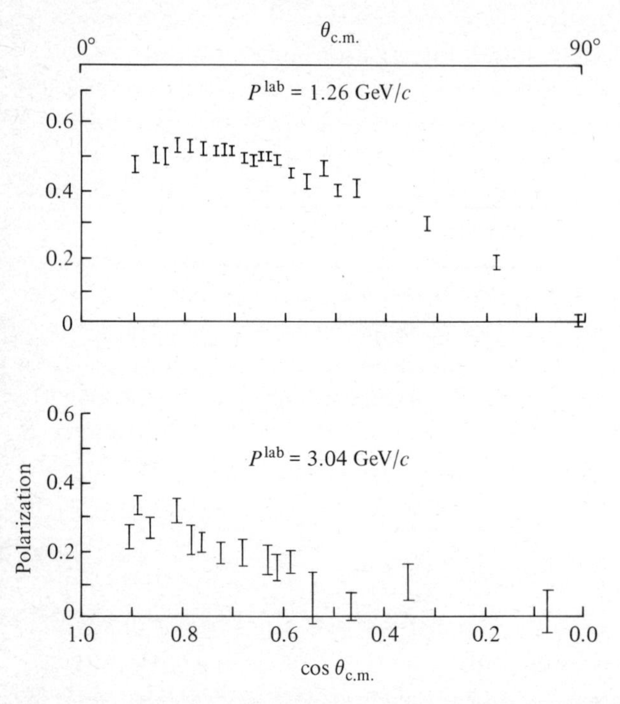

Figure 3-11 Typical experimental data for the polarization in p-p elastic scattering as a function of cos $\theta_{c.m.}$. (*From UCRL,* 20,000 *NN.*)

these circumstances the electron magnetic moments in the solid can achieve quite large polarization since $\mu_e \gg \mu_p$. The magnetic moments of the electrons and protons are coupled through their magnetic interaction (Prob. 3-17). The protons can therefore be polarized indirectly by saturating the crystal with microwaves of a frequency appropriate for flipping the spin of the protons and electrons together.[1] Proton polarizations of 60 percent can regularly be obtained (Prob. 3-5). Experiments which use such targets must incorporate a means of distinguishing scatterings off of free protons from those off of protons bound in nuclei (which are unpolarized). This can usually be done by requiring that the events satisfy the

[1] The discussion given here is oversimplified. The details are rather complicated. For a complete discussion, see C. D. Jeffries, "Dynamic Nuclear Orientation," Interscience–Wiley, New York, 1963.

kinematic constraints imposed by conservation of energy and momentum.

Figure 3-11 shows some typical experimental data for the polarization in *p-p* scattering. In addition to the polarization measurement described above, other more complicated experiments to study the spin dependence in *N-N* scattering can be carried out. These typically use three successive scatterings (if a polarized target is not used), with one scattering plane at right angles to the other two.[1]

THE YUKAWA HYPOTHESIS

One aspect of the force between nucleons that is understood reasonably well is the long-range behavior. Early studies of nuclei showed that the range of the *N-N* interaction is approximately 1.4×10^{-13} cm; in other words, the force falls away rapidly beyond that distance. This is in contrast with the coulomb force which varies as $1/r^2$ and has no characteristic length associated with it. In 1935 Yukawa proposed that this type of behavior could be explained if the *N-N* force was due to the exchange of quanta of nonzero mass.

To understand how this comes about, let us first consider the electromagnetic force. The electromagnetic field is described classically by Maxwell's equations. To describe quantum behavior, it is necessary to quantize the field. This is the subject of quantum electrodynamics, which cannot be considered here in any detail. Qualitatively we can say that the electromagnetic fields are replaced by a cloud of photons with the energy of the field equal to the sum of the energies of the photons. The force between two charged particles is then ascribed to the exchange of photons. The collision between an electron and a proton can be schematically represented as shown in Fig. 3-12*a*. The electron is thought of as emitting a photon which is absorbed by the proton, or vice versa. In the process, momentum is transferred from the electron to the proton. The electron in emitting the photon violates energy conservation. However this is allowed by the uncertainty principle provided the photon only exists for a time Δt such that $\Delta E \, \Delta t \lesssim \hbar$, where ΔE is the energy of the photon. A process that occurs only by virtue of such an energy violation is called a *virtual process*, and the exchanged photon is said to be a *virtual photon*.

[1] See, for example, E. G. Segrè, "Nuclei and Particles," Benjamin, New York, 1964.

Hideki Yukawa. Born 1907 in Tokyo. He received a D.Sc. from Osaka University in 1938. In 1935 he proposed that the force responsible for nuclear binding was a new type of force, neither electromagnetic nor gravitational in origin. To account for the finite range of this force, he postulated that there must be associated with it a new particle with a mass approximately 200 times that of the electron. The π meson, whose existence was confirmed in 1947, proved to be the "Yukawa particle." Yukawa was awarded the Nobel Prize in 1949 for his prediction of the existence of mesons. (*Photograph from the Niels Bohr Library, American Institute of Physics.*)

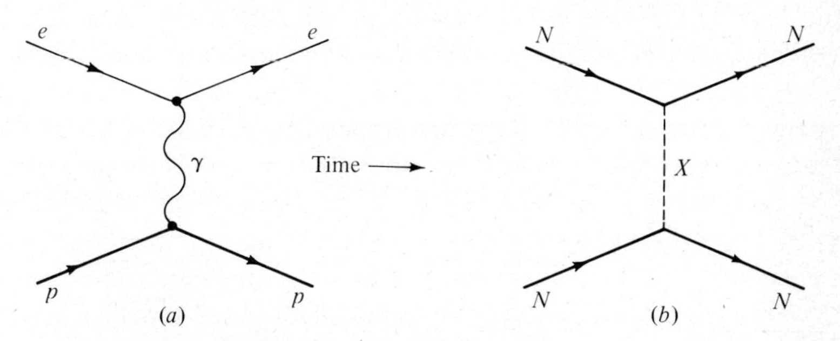

Figure 3-12 (*a*) Schematic representation of an elastic *e-p* scattering. The vertical axis is the spatial separation; the horizontal axis is time. Momentum is transferred from one particle to the other through the exchange of a photon. (*b*) Elastic nucleon-nucleon scattering through the exchange of a particle X with nonzero rest mass.

We shall return to the diagram in Fig. 3-12*a* later when we discuss *e-p* scattering. We now want to consider Yukawa's hypothesis for the N-N force. By analogy with *e-p* scattering, we imagine N-N scattering as due to the exchange of a particle, say X, with a nonzero rest mass (Fig. 3-12*b*). We can estimate the mass of the exchanged particle from the range of the N-N interaction by a simple argument: The energy violation ΔE associated with the emission of the particle is $\sim m_X c^2$. From the uncertainty principle the particle can only exist for a time

$$\Delta t \lesssim \frac{\hbar}{\Delta E} \approx \frac{\hbar}{m_X c^2}$$

In this time it can travel, at most, a distance $c\,\Delta t$, where c is the velocity of light. If we equate this with the range of the N-N interaction, we obtain a rough estimate of the mass of the exchanged particle,[1]

$$m_X c^2 \approx \frac{\hbar c}{1.4 \times 10^{-13}\ \text{cm}} \approx 140\ \text{MeV}$$

(Yukawa's prediction was actually 200 electron masses = 102 MeV.)

[1] In units convenient for high-energy physics problems $\hbar c = 197.3$ MeV-fm. A compilation of useful constants appears in Appendix A.

If such particles exist virtually in the vicinity of nucleons, it should be possible to "shake them loose" by supplying enough energy to the nucleon (just as photons are shaken loose in bremsstrahlung). Shortly after Yukawa made his proposal, a particle of about the right mass was discovered in cosmic radiation. This turned out to be the muon ($mc^2 = 105.6$ MeV). Before long it was realized that the interaction between the muon and the nucleon was much too weak to account for the strong force between nucleons. This puzzle was eventually resolved when the $\pi^{\pm}$ and later the π^0 were discovered, with rest masses $\sim$140 MeV. These interacted strongly with nucleons as the Yukawa particle must.

In analogy with the picture of the electromagnetic force given previously, one might picture the nucleon as being surrounded by a cloud of virtual pions. There are in addition, however, many possible virtual particles other than pions which the nucleon can emit. These include η mesons, K-meson pairs, nucleon-antinucleon pairs, etc. These all contribute to the interaction between nucleons, but one can conclude from the argument above that their contributions will have a shorter range than the one-pion-exchange contribution because of the larger masses involved. Thus one-pion exchange dominates the long-range behavior of the N-N interaction. This contribution can be calculated fairly reliably, but the short-range behavior seems hopelessly complicated at the present time.

INELASTIC SCATTERING

For sufficiently high energies a wide variety of inelastic processes can take place in N-N scattering, for example,

$$\begin{aligned} N + N &\rightarrow N + N + m\pi \qquad m = 1, 2, 3, \ldots \\ &\rightarrow N + N + K + \bar{K} \\ &\rightarrow N + \Lambda^0 + K \\ &\rightarrow N + N + \bar{N} + N \\ &\rightarrow \cdots \end{aligned}$$

Reactions such as these are of great practical importance as sources of pion, kaon, and antinucleon beams from proton accelerators.[1] The analysis of such reactions is quite difficult from both an experi-

[1] The target nucleon may be bound in a nucleus.

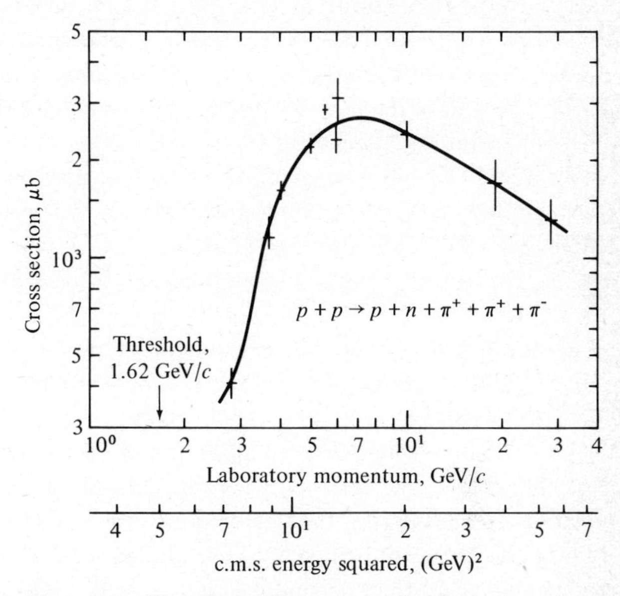

Figure 3-13 Variation of the cross section for $p + p \rightarrow p + n + 2\pi^+ + \pi^-$ with laboratory momentum. (*From UCRL, 20,000 NN.*)

mental and a theoretical point of view, and we shall not discuss such reactions in detail.

One interesting generalization we can make concerns the energy dependence of the cross section for any given inelastic process. Experimentally the total cross section remains roughly constant at high energies (Figs. 3-4 and 3-5). As the total energy in the c.m.s. is increased, more and more inelastic channels open up. Each, in a sense, competes for its share of the total cross section, so that at high energies the cross section for a given channel generally decreases with increasing energy. Since there is usually a rapid rise in the cross section just beyond threshold, the energy dependence of the cross section for a typical inelastic process is like that shown in Fig. 3-13.

3-4 PHASE-SHIFT ANALYSIS OF SCATTERING

In order to compare the results from scattering experiments more readily with theory and as a convenient means of parameterizing these results, it is useful, at least at low and moderate energies, to make phase-shift analyses of the data. This involves expressing the amplitude of the scattered wave in terms of quantities called *phase shifts,* which describe how different angular-momentum states of the incident beam are affected in the scattering.

To avoid becoming bogged down in complexity, we shall consider the simplest possible case, that of elastic scattering of spinless particles. We shall also use nonrelativistic quantum mechanics, although the final results in fact turn out to be applicable in relativistic situations.[1]

In the c.m.s. of the incident and target particles, the two-body wave equation can be reduced to an equivalent one-body equation as for the hydrogen atom. It is convenient to represent the incident beam as a plane wave directed along the $+Z$ axis. (This is equivalent, because of the uncertainty principle relating momentum and position, to considering the vector momentum of the incident beam to be perfectly well defined.) The problem can be treated as time independent and the t dependence suppressed. The incoming wave is then written as e^{ikz}, where $k = p/\hbar$. The asymptotic form (at large distances from the point of interaction) of the wave function for the system will consist of the incident plane wave plus an outgoing spherical wave,

$$\psi(r,\theta) = e^{ikz} + f(\theta)\,\frac{e^{ikr}}{r} \tag{3-5}$$

The intensity of the outgoing wave is obtained from the absolute square of the last term, $|f(\theta)|^2/r^2$. The differential elastic-scattering cross section will therefore be proportional to $|f(\theta)|^2$. In fact, with the normalization for ψ chosen in Eq. (3-5), it turns out that

$$\frac{d\sigma}{d\Omega} = |f(\theta)|^2 \tag{3-6}$$

The wave function $\psi(r,\theta)$ can also be expressed as a linear combination of eigenfunctions of the Schrödinger equation with a central

[1] We give only an outline of the treatment. A more rigorous and complete discussion is given in many quantum mechanics texts. See, for example, L. I. Schiff, "Quantum Mechanics," 3d ed., McGraw-Hill, New York, 1968. For a more general treatment including spin, see B. T. Feld, "Models of Elementary Particles," app. 5, Blaisdell, Waltham, Mass., 1969.

force,[1]

$$\psi(r,\theta) = \sum_{\ell} R_\ell(r) P_\ell(\cos\theta)$$

The $P_\ell(\cos\theta)$ are Legendre polynomials of order ℓ, and $R_\ell(r)$ gives the radial dependence. The summation index ℓ is identified with the orbital angular-momentum quantum number.

The functions $R_\ell(r)$ can in principle be determined from a knowledge of the force acting between the particles. In practice this is usually not known a priori. For large r, well outside the range of the force, $R_\ell(r)$ must have the form of a spherical outgoing wave with the same wavelength and wave number k as the incident beam; therefore asymptotically

$$R_\ell(r) \propto \frac{1}{r} \sin\left(kr - \tfrac{1}{2}\ell\pi + \delta_\ell\right)$$

The δ_ℓ are the *phase shifts* for partial waves of angular momentum ℓ. They are defined in such a way that $\delta_\ell \to 0$ if there is no interaction.

A relation between $f(\theta)$ and the phase shifts can be obtained by expanding the incident wave $e^{ikz} = e^{ikr\cos\theta}$ in a series of Legendre polynomials, substituting this in Eq. (3-5), and equating the two asymptotic expressions for $\psi(r,\theta)$. The result is[2]

$$f(\theta) = \frac{1}{2ik} \sum_{\ell} (2\ell + 1)(e^{2i\delta_\ell} - 1) P_\ell(\cos\theta) \qquad (3\text{-}7)$$

If the force between the particles can be described by a potential function $U(r)$, the δ_ℓ can be calculated by solving the radial part of the wave equation, either analytically or numerically. In practice,

[1] We are considering the scattering of spinless particles, so that in the absence of any external fields the forces acting between the particles and the wave function ψ must be independent of the polar angle ϕ.

[2] Note that any function $f(\theta)$ can be expanded as a sum of Legendre polynomials. The only physics here is the identification of the δ_ℓ with the phase shift of that part of the incident wave with angular momentum ℓ. The physics comes when we try to relate the δ_ℓ to the forces acting between the two particles.

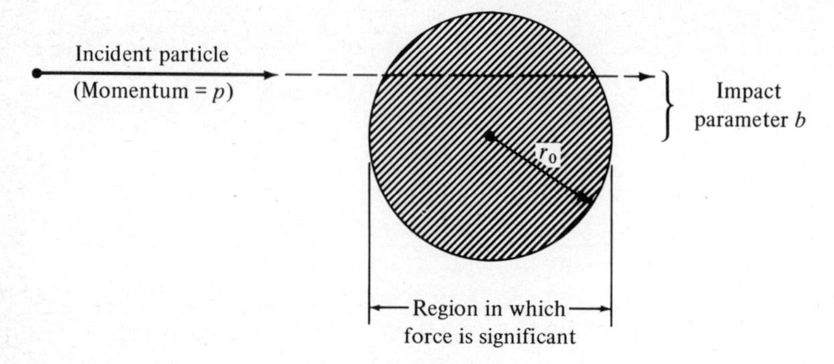

Figure 3-14 Classical picture of a beam particle passing near a force center.

however, one usually knows $d\sigma/d\Omega$, which is experimentally accessible, and wants to determine $U(r)$. There is no general procedure for doing this. Furthermore it is found that, especially for N-N scattering, the form of the potential required to fit the data, even at relatively low energies (below the π-meson production threshold), is quite complicated and of doubtful usefulness.[1] The usual practice is therefore to work directly with the δ_ℓ, which at least provide a concise parameterization of the scattering data.

At low and moderate energies, only the first few angular-momentum states need be considered in the scattering amplitude. The number required can be estimated by a semiclassical argument. All forces that we shall consider, except the coulomb force, are characterized by a rather well-defined range r_0, such that for $r > r_0$, the force is negligible. In the case of the N-N interaction, $r_0 \approx 1.4$ fm and is determined by the mass of the pion (see Sec. 3-3, the Yukawa Hypothesis). Classically the angular momentum of a particle is related to the impact parameter b (Fig. 3-14) through the relation

$$L = pb$$

Particles with $b > r_0$ will not be scattered appreciably. Quantum mechanically the angular momentum is quantized so that $L = \sqrt{\ell(\ell + 1)}\,\hbar \approx \ell\hbar$. The maximum ℓ that will contribute in

[1] In general the potential can depend on spin orientation, angular momentum, and energy. See, for example, Segrè, *op. cit.*, sec. 10-6.

Eq. (3-7) will therefore be given by

$$\hbar \ell_{\max} \approx p r_0 \tag{3-8}$$

For $r_0 = 1.4$ fm and p in MeV/c, $\ell_{\max} \approx p/140$.[1] At very low energies $\ell_{\max} \sim 0$, and from Eqs. (3-6) and (3-7)

$$\begin{aligned}\frac{d\sigma}{d\Omega} &= |f(\theta)|^2 \\ &= \frac{1}{4k^2}\,|e^{2i\delta_0} - 1|^2\end{aligned}$$

The scattering is therefore isotropic. In language borrowed from atomic physics, $\ell = 0$ scattering is referred to as *s-wave scattering*, $\ell = 1$ as *p wave*, etc.

The total elastic cross section can be found by integrating $|f(\theta)|^2$ over a sphere with the use of the orthogonality of the Legendre polynomials. The result is

$$\sigma_T = 4\pi\lambda\!\!\!^{-2} \sum (2\ell + 1)\sin^2\delta_\ell \tag{3-9}$$

where $\lambda\!\!\!^{-} = \hbar/p = k^{-1}$. The maximum contribution from a given ℓ is therefore

$$\sigma_\ell^{\max} = (2\ell + 1)4\pi\lambda\!\!\!^{-2} \tag{3-9a}$$

for $\delta_\ell = 90°$. When a given δ_ℓ becomes equal to 90°, that angular-momentum state is said to be in *resonance*. This will often cause a prominent peak in a plot of the cross section versus momentum, with the maximum at or near the momentum at which $\delta_\ell = 90°$. (See, for example, the π-p system discussed in the next section.) Often the resonant amplitude dominates the total cross section so that the peak value of σ_T gives a good indication of which angular-momentum state is resonant.

An important result can be derived from the expression for $f(\theta)$ above by setting $\theta = 0$. We have

$$f(0) = \frac{1}{2ik}\sum\,[(2\ell + 1)(\cos 2\delta_\ell + i\sin 2\delta_\ell - 1)\cdot 1]$$

[1] Here p is the momentum in the c.m.s. of the two particles.

so that if Im $f(0)$ is the imaginary part of the forward scattering amplitude,

$$\begin{aligned}
\operatorname{Im} f(0) &= -\frac{1}{2k}\sum (2\ell+1)(\cos 2\delta_\ell - 1) \\
&= \lambda\!\!\!^{-} \sum (2\ell+1)\sin^2\delta_\ell \\
&= \frac{\sigma_T}{4\pi\lambda\!\!\!^{-}}
\end{aligned} \tag{3-10}$$

This relation is called the *optical theorem.* It turns out to be a very general theorem and works even in situations where inelastic processes (discussed below) can occur.

The cross section at 0° is

$$\begin{aligned}
\frac{d\sigma}{d\Omega}(0^\circ) &= |f(0)|^2 \\
&= [\operatorname{Re} f(0)]^2 + [\operatorname{Im} f(0)]^2
\end{aligned}$$

Typically Im $f(0) \gg$ Re $f(0)$ at high energies, and since σ_T is roughly constant at high energies, we can conclude

$$\begin{aligned}
\frac{d\sigma}{d\Omega}(0^\circ) &\propto \frac{1}{\lambda\!\!\!^{-2}} \\
&\propto p^2
\end{aligned}$$

for large p, where p is the momentum in the c.m.s.

Except for the generalized optical theorem discussed above, the previous results hold only for elastic scattering. Inelastic processes, such as pion production, correspond to absorption of the incident wave (analogous to the scattering of light from a cloudy water droplet). The above formulas can readily be generalized to include absorption. The only difference is that the δ_ℓ become complex parameters. Rather than deal with complex exponentials, it is more convenient to define

$$\begin{aligned}
\eta_\ell &\equiv e^{2i\delta_\ell} \\
&= e^{2i\delta_\ell{}^R} e^{-2\delta_\ell{}^I} \\
&= |\eta_\ell| e^{2i\delta_\ell{}^R}
\end{aligned}$$

We can now drop the superscript on $\delta_\ell{}^R$ with the understanding that δ_ℓ is real. The resulting expressions for the scattering amplitude and the elastic, inelastic, and total cross sections are then

$$\begin{aligned} f(\theta) &= \frac{1}{2ik}\sum_{\ell}(2\ell+1)(|\eta_\ell|e^{2i\delta_\ell}-1)P_\ell(\cos\theta) \\ \sigma_{\text{el}} &= \pi\lambda\!\!\!^{-2}\sum(2\ell+1)|1-\eta_\ell|^2 \\ \sigma_{\text{inel}} &= \pi\lambda\!\!\!^{-2}\sum(2\ell+1)(1-|\eta_\ell|^2) \\ \sigma_T &= \sigma_{\text{el}}+\sigma_{\text{inel}} \\ &= 2\pi\lambda\!\!\!^{-2}\sum(2\ell+1)(1-\operatorname{Re}\eta_\ell) \end{aligned} \tag{3-11}$$

Note that no absorption corresponds to $|\eta_\ell| = 1$, and complete absorption to $|\eta_\ell| = 0$. Even in the latter case there is a contribution to σ_{el}. In fact for $|\eta_\ell| = 0$,

$$\begin{aligned} \sigma_{\text{el}}{}^{\ell} &= \sigma_{\text{inel}}^{\ell} \\ &= \pi\lambda\!\!\!^{-2}(2\ell+1) \end{aligned}$$

The elastic scattering resulting from absorption is just the diffraction scattering referred to previously. In the limit of large ℓ_{max}, with total absorption for $\ell < \ell_{\text{max}}$ and no absorption for $\ell > \ell_{\text{max}}$ (scattering from a "black disk"),

$$\begin{aligned} \sigma_T &= 2\pi\lambda\!\!\!^{-2}\sum_{\ell=0}^{\ell_{\text{max}}}(2\ell+1) \to 2\pi\lambda\!\!\!^{-2}\int_0^{\ell_{\text{max}}}(2\ell+1)\,d\ell \\ &\approx 2\pi R^2 \end{aligned}$$

where $\ell_{\text{max}} = pR/\hbar = R/\lambda\!\!\!^{-}$, with R the radius of the disk. This result that the total cross section is twice the geometric cross section is surprising to one used to thinking of particles as being like machine-gun bullets. In such a picture σ_T would be expected to be equal to $\pi R^2 = \sigma_{\text{inel}}$. However because of the *diffraction* of the wave around the target, we have to include in σ_T an *elastic* contribution, which for a black disk (that is, $|\eta_\ell| = 0$ for $\ell < \ell_{\text{max}}$) turns out to equal the inelastic cross section. The elastic scattering is sharply peaked in the forward direction with a characteristic width in angle of about λ/R, just as in the diffraction of light around a circular obstacle.

Phase-shift analysis becomes considerably more complicated if either or both particles have spin. In the simplest case where one particle has spin $\frac{1}{2}$ (for example, π-N scattering), the scattering amplitude can depend on the relative orientation of the spin before and after scattering. There are then two amplitudes, usually referred to as the *spin-flip amplitude* and the *nonflip amplitude*, each given by a formula like Eq. (3-7). Therefore there are effectively twice as many parameters (phase shifts) needed to describe the scattering at a given energy. If both particles have spin $\frac{1}{2}$, as in N-N scattering, five amplitudes are needed to describe the scattering.

Despite the complexity of the problem, rather complete phase-shift analyses have been made for π-N scattering with laboratory energies up to approximately 2.5 GeV and for N-N scattering up to approximately 1 GeV. These analyses require quite elaborate computer programs which make "fits" to all the available scattering and polarization data. A smooth dependence of the phase shifts on energy is invariably assumed.

3-5 THE π-N INTERACTION

ISOTOPIC SPIN CONSIDERATIONS

The π-N system turns out to be considerably richer than the N-N system, and an enormous effort has gone into its study. Superficially one might expect that it would be necessary to study experimentally the scattering of π^+, π^-, and π^0 mesons on neutrons and protons to specify the π-N interaction completely. However isotopic spin considerations simplify the problem. The total isospin vector $\mathbf{T}$ for the π-N system is $\mathbf{T} = \mathbf{T}_\pi + \mathbf{T}_N$. The pion has $T_\pi = 1$, and the nucleon $T_N = \frac{1}{2}$. The vector $\mathbf{T}_\pi$ can be either parallel or antiparallel to $\mathbf{T}_N$, so the only possible values of T are $\frac{1}{2}$ or $\frac{3}{2}$. Any π-N system can therefore be regarded as a mixture of $T = \frac{1}{2}$ and $T = \frac{3}{2}$ states as far as strong interactions are concerned.

The π^+-p system has $T_3 = Q/e - B/2 = \frac{3}{2}$, and therefore only the $T = \frac{3}{2}$ amplitude contributes in π^+-p scattering. Similarly π^--n has $T_3 = -\frac{3}{2}$, so it is a pure $T = \frac{3}{2}$ system and equivalent to the π^+-p system. The π^--p system has $T_3 = \frac{1}{2}$. We can get $T_3 = \frac{1}{2}$ either from a T vector with $T = \frac{1}{2}$ directed along the T_3 axis in isospin space or from a T vector with $T = \frac{3}{2}$ at an appropriate angle to the T_3 axis. Therefore the π^--p system must be considered a mixture of $T = \frac{1}{2}$ and $T = \frac{3}{2}$ states (in this case, however, not an equal mixture). Similarly the π^+-n, the π^0-p, and the π^0-n systems are also linear combinations of $T = \frac{1}{2}$ and $T = \frac{3}{2}$ states. There are straightforward techniques in quantum mechanics for constructing the appropriate linear combinations of $T = \frac{1}{2}$ and $T = \frac{3}{2}$ for each case; however we shall just quote the results. (The problem is completely analogous to combining ordinary spins.)

Isotopic spin invariance leads to relations between cross sections for the various π-N systems. For the total cross sections,

$$\begin{aligned} \sigma_T(\pi^+\text{-}p) &= \sigma_T(\pi^-\text{-}n) = \sigma_T^{(\frac{3}{2})} \\ \sigma_T(\pi^-\text{-}p) &= \sigma_T(\pi^+\text{-}n) = \tfrac{2}{3}\sigma_T^{(\frac{1}{2})} + \tfrac{1}{3}\sigma_T^{(\frac{3}{2})} \\ \sigma_T(\pi^0\text{-}p) &= \sigma_T(\pi^0\text{-}n) = \tfrac{1}{3}\sigma_T^{(\frac{1}{2})} + \tfrac{2}{3}\sigma_T^{(\frac{3}{2})} \end{aligned} \tag{3-12}$$

Here $\sigma_T^{(\frac{3}{2})}$ and $\sigma_T^{(\frac{1}{2})}$ are total cross sections for the pure isospin states. From the experimentally accessible π^+-p and π^--p cross sections, we can therefore predict all the others, in particular, the experimentally inaccessible π^0-n and π^0-p cross sections.

Similar relations hold for elastic scattering. If we write $d\sigma/d\Omega = |f|^2$ and denote the scattering amplitudes for the pure isospin states as $f_{\frac{1}{2}}$ and $f_{\frac{3}{2}}$,[1] then

$$\begin{aligned} f(\pi^+ + p \to \pi^+ + p) &= f_{\frac{3}{2}} \\ f(\pi^- + p \to \pi^- + p) &= \tfrac{2}{3}f_{\frac{1}{2}} + \tfrac{1}{3}f_{\frac{3}{2}} \\ f(\pi^0 + p \to \pi^0 + p) &= \tfrac{1}{3}f_{\frac{1}{2}} + \tfrac{2}{3}f_{\frac{3}{2}} \end{aligned} \tag{3-13}$$

A particularly interesting result is

$$\begin{aligned} f(\pi^- + p \to \pi^0 + n) &= f(\pi^+ + n \to \pi^0 + p) \\ &= \frac{-\sqrt{2}}{3} f_{\frac{1}{2}} + \frac{\sqrt{2}}{3} f_{\frac{3}{2}} \end{aligned} \tag{3-13a}$$

The reactions $\pi^- + p \to \pi^0 + n$ and $\pi^+ + n \to \pi^0 + p$ are referred to as charge-exchange scattering, analogous to n-p charge-exchange scattering, discussed in Sec. 3-3.

REVIEW OF EXPERIMENTAL RESULTS

The total cross sections for π^+-p and π^--p scattering are shown in Fig. 3-15. It is clear that these show considerably more structure

[1] As discussed in Sec. 3-4, two amplitudes are needed to describe π-N scattering, the spin-flip and nonflip amplitudes. Equations (3-13) hold for each separately. Each of the amplitudes is, of course, complex.

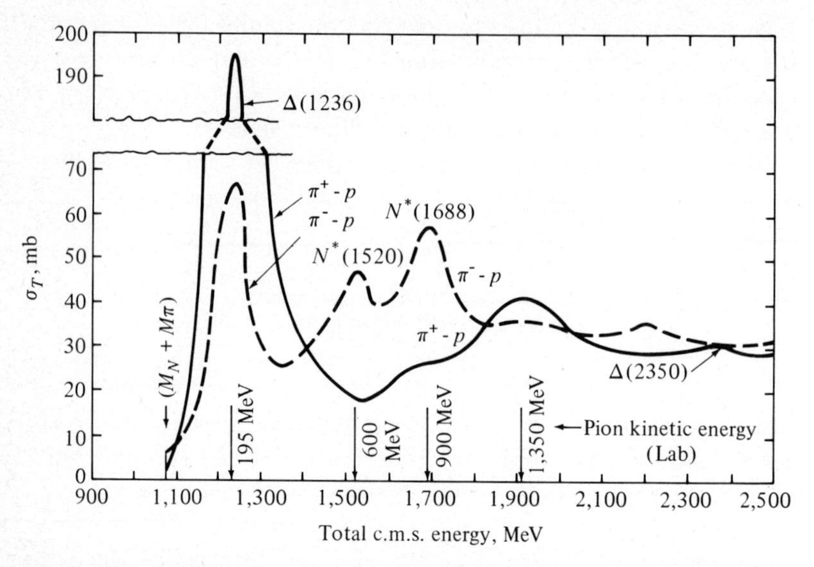

Figure 3-15 Total π^+-p and π^--p cross sections versus energy in the c.m.s. (*Based on a compilation by the Particle Data Group.*)

than the N-N total cross sections. The most striking feature is the large peak at about 195 MeV (pion kinetic energy in the laboratory), which is more prominent in π^+-p scattering. Such peaks are common in nuclear physics and are often referred to as resonances.[1] In atomic and nuclear physics these occur when the energy of the incident particle is just right to excite some higher energy level in the atom or nucleus. This peak in the π-N system corresponds to the formation of an excited state of the nucleon through the reaction

$$\pi^+ + p \to N^{*++}$$

where the N^* decays very quickly. At these energies the most likely decay mode is $N^{*++} \to \pi^+ + p$. The overall reaction is shown schematically in Fig. 3-16. From the observed position of the peak,

[1] Strictly speaking, for a true resonance a dominant phase shift should pass through 90° at or near the peak of the resonance (Sec. 3-4). This is true for π-p scattering near 200 MeV. However the name *resonance* is now given generically to any fairly sharp bump in a total cross section, even though a detailed phase-shift analysis has not been made.

the mass of the N^* is approximately 1,236 MeV, the total energy in the c.m.s. at the peak. The width of the peak in energy is directly related to the lifetime of the N^* through the uncertainty principle. From Fig. 3-15, $\Delta E \approx 100$ MeV, so

$$\begin{aligned}\Delta t &\approx \frac{\hbar}{\Delta E} \\ &= \frac{6.6 \times 10^{-22}\ \text{MeV-sec}}{100\ \text{MeV}} \\ &= 6.6 \times 10^{-24}\ \text{sec}\end{aligned}$$

This is indeed a short time, comparable to the time it takes a relativistic particle to cross the nucleon! It is clear the N^* decays through a strong interaction since typical electromagnetic and weak decays are orders of magnitude slower (Table 1-1).

Despite the short lifetime of the N^*, most high-energy physicists now believe that such resonances have just as much right to be considered particles as the more stable states listed in Table 1-1. In view of our lack of any basic understanding of the strong interactions, we must consider it only an "accident" that the mass of the N^* is not 170 MeV lower. If it were, it could only decay through an electromagnetic interaction into a nucleon plus either leptons or γ's, and its lifetime would be comparable to the π^0, η^0, or Σ^0. Historically the $N^*(1236)$, or $\Delta(1236)$ as it is more properly called,[1] was the first of many such short-lived states to be discovered. Two others, the $N^*(1520)$ and $N^*(1688)$, also show up as rather well-

[1] The symbol Δ is used for states with isospin $\frac{3}{2}$ to distinguish them from states with $T = \frac{1}{2}$ such as the nucleon.

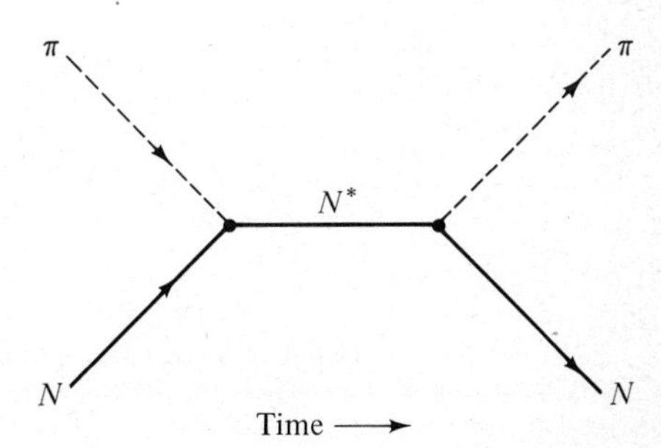

Figure 3-16 The absorption of a pion by a nucleon to form an N^*, which subsequently decays back into a pion and nucleon.

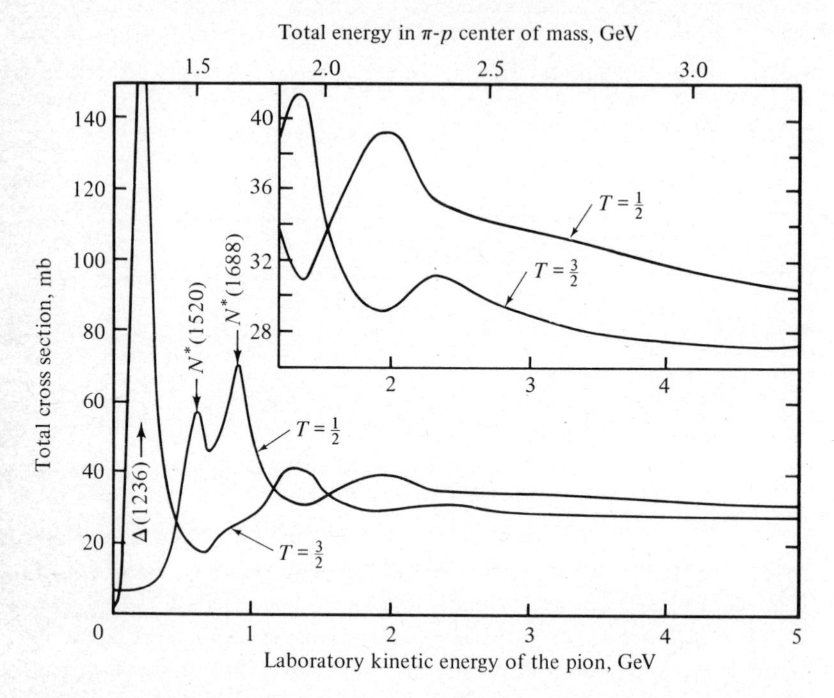

Figure 3-17 Total cross sections for pure isospin $\frac{1}{2}$ and $\frac{3}{2}$ π-N states versus kinetic energy of the pion in the laboratory. The inset at the upper right shows the higher-energy portion on a magnified scale. *(Based on a compilation by the Particle Data Group.)*

defined peaks in the π^--p total cross section, at about 600 and 900 MeV respectively.[1] Such states are often referred to as *nucleon isobars.* Analogous short-lived states are also found in the π-π system (Sec. 3-7) and most other systems.

It is of considerable interest to establish the quantum numbers of these states. The isotopic spin assignments become obvious if we plot the total cross sections for the pure isospin states which can be determined from Eqs. (3-12). These are plotted in Fig. 3-17. It is apparent that the $\Delta(1236)$ has $T = \frac{3}{2}$ and the $N^*(1520)$ and $N^*(1688)$ both have $T = \frac{1}{2}$.

[1] Several others exist as well, but these only appear clearly in detailed phase-shift analyses of π-N scattering, where they show up as phase shifts passing through 90°.

The spin (or intrinsic angular momentum) J of a resonance can often be determined by studying the angular distribution of its decay products, or in this case, the angular distribution in elastic π-N scattering. If the resonant term dominates the elastic scattering amplitude, then at the peak of the resonance the angular distribution takes on a characteristic shape.[1] If a complete phase-shift analysis is possible, the spin of the resonant state is easily obtained as the total angular momentum of the phase shift which passes through 90°.

The spin of the $\Delta(1236)$ is readily determined from qualitative arguments. The resonance appears to dominate the scattering cross section near 200 MeV. The maximum contribution to the cross section from any angular-momentum state is

$$\sigma_J^{\max} = (2J + 1)2\pi\lambda\!\!\!^{-2} \tag{3-14}$$

where J is the total angular momentum.[2] Since the nucleon has spin $\frac{1}{2}$ and the pion has spin 0, then $J = \ell \pm \frac{1}{2}$ and is always half-integer. Setting $(2J + 1)2\pi\lambda\!\!\!^{-2} \approx 200$ mb, the value of the π^+-p cross section at the peak of the resonance, we find that $J = \frac{3}{2}$ is the only possibility (Prob. 3-14). The orbital angular momentum ℓ can be $J + \frac{1}{2}$ or $J - \frac{1}{2}$. The semiclassical relation, Eq. (3-8), with $r_0 \approx 1$ fm gives $\ell_{\max} \approx 1$, which suggests $\ell = 1$ for the resonance. This is verified by other evidence. A state with $\ell = 1$ has odd orbital parity, but because of the odd intrinsic parity of the pion, the overall parity will be even. In the notation adopted in Sec. 1-3, the spin-parity assignment for the $\Delta(1236)$ is therefore $\frac{3}{2}^+$. Probable assignments for the $N^*(1520)$ and $N^*(1688)$ are $J = \frac{3}{2}^-$ and $J = \frac{5}{2}^+$ respectively. It is an interesting empirical fact that higher mass resonances tend to be associated with higher spins (Table 3-1).

[1] For spinless particles, for example, we should expect, from Eq. (3-7) with $\delta_l = 90°$, that $d\sigma/d\Omega \propto [P_l(\cos\theta)]^2$, where $J = \ell$ is the angular momentum of the resonant state.

[2] This comes from the appropriate generalization of Eq. (3-9a) given previously for spinless particles. If there is negligible absorption,

$$\sigma_J^{\max} = 4\pi\lambda\!\!\!^{-2}\,\frac{2J + 1}{(2j_1 + 1)(2j_2 + 1)}$$

where j_1 and j_2 are the spins of the incident and target particles.

Table 3-1 Examples of particles which decay through strong interactions
Doubtful assignments indicated by (?)

Symbol	*S*	T^G	J^P	*Mass, MeV*	*Width, MeV*	*Decay modes* *Mode*	*Fraction %*
I. Mesons							
$\rho(765)$	0	1^+	1^-	765 ± 10	≈ 125	$\pi\pi$	≈ 100
						e^+e^-	0.006
						$\mu^+\mu^-$	0.007
$\omega(784)$	0	0^-	1^-	783.9 ± 0.3	11	$\pi^+\pi^-\pi^0$	90
						$\pi^+\pi^-$	1
						$\pi^0\gamma$	9
$\eta'(958)$ or X^0	0	0^+	0^- ?	957.5 ± 0.8	<4	$\eta\pi\pi$	64
						$\pi^+\pi^-\gamma$	29
						$\gamma\gamma$	7
$\phi(1019)$	0	0^-	1^-	$1{,}019.5 \pm 0.6$	4	K^+K^-	47
						$K_L{}^0K_S{}^0$	35
						$\pi^+\pi^-\pi^0$	18
$B(1235)$	0	1^+	1^+ ?	$1{,}233 \pm 10$	≈ 100	$\omega\pi$	≈ 100
$f(1260)$	0	0^+	2^+	$1{,}269 \pm 10$	≈ 150	$\pi\pi$	≈ 80
$K^*(892)$	+1 (K^{*+})	$\frac{1}{2}$	1^-	893 ± 0.5	50	$K\pi$	≈ 100
						$K\pi\pi$	0.2
$K_N(1420)$	+1 ($K_N{}^+$)	$\frac{1}{2}$	2^+ ?	$1{,}408 \pm 10$	≈ 107	$K\pi$	57
						$K^*\pi$	27
						$K\rho$	9

Typical low-energy angular distributions for π^+-p scattering are shown in Fig. 3-18. At the peak of the resonance it is found that

$$\frac{d\sigma}{d\Omega} \approx \lambda\!\!\!^{-2}(1 + 3\cos^2\theta)$$

This is exactly the result expected from a phase-shift analysis if the ($T = \frac{3}{2}$, $J = \frac{3}{2}$) amplitude dominates and δ_{33} is equal to 90° [the subscript is $(2T,2J)$].

At higher energies as inelastic processes become increasingly important, the forward diffraction peak dominates the angular distribution (Fig. 3-19).

Table 3-1 Examples of particles which decay through strong interactions *(Continued)*

Symbol	S	T^G	J^P	Mass, MeV	Width, MeV	Decay modes: Mode	Decay modes: Fraction %
II. Baryons							
$\Delta(1236)$	0	$\frac{3}{2}$	$\frac{3}{2}^+$	1,230–1,236	≈ 120	$N\pi$	99.4
$\Delta(1950)$	0	$\frac{3}{2}$	$\frac{7}{2}^+$	1,930–1,980	140–220	$N\pi$	45
						$\Delta(1236)\pi$	≈ 50
						ΣK	2
						etc.	
$N^*(1520)$	0	$\frac{1}{2}$	$\frac{3}{2}^-$	1,510–1,540	105–150	$N\pi$	50
						$N\pi\pi$	50
$N^*(1670)$	0	$\frac{1}{2}$	$\frac{5}{2}^-$	1,655–1,680	105–175	$N\pi$	40
						$N\pi\pi$	60
$N^*(1688)$	0	$\frac{1}{2}$	$\frac{5}{2}^+$	1,680–1,692	105–180	$N\pi$	60
						$N\pi\pi$	40
$N^*(2190)$	0	$\frac{1}{2}$	$\frac{7}{2}^-$	2,000–2,260	~ 300	$N\pi$	25
						$N\pi\pi$	(observed)
$\Lambda(1405)$	-1	0	$\frac{1}{2}^-$	$1,405 \pm 5$	40 ± 10	$\Sigma\pi$	100
$\Lambda'(1520)$	-1	0	$\frac{3}{2}^-$	$1,518 \pm 2$	16 ± 2	$N\bar{K}$	46
						$\Sigma\pi$	41
						etc.	
$\Sigma(1385)$	-1	1	$\frac{3}{2}^+$	1,383	36 ± 3	$\Lambda\pi$	90
						$\Sigma\pi$	10
$\Sigma(1765)$	-1	1	$\frac{5}{2}^-$	$1,765 \pm 5$	~ 120	$N\bar{K}$	44
						$\Lambda\pi$	15
						$\Lambda'(1520)\pi$	14
						etc.	
$\Xi(1530)$	-2	$\frac{1}{2}$	$\frac{3}{2}^+$	1,529	7.3 ± 1.7	$\Xi\pi$	100

SOURCE: Data based on a compilation by the Particle Data Group.

The spin dependence of the π-N interaction can be studied in scattering experiments with polarized targets.

Another interesting class of experiments possible with pion beams is the production of kaons and hyperons through reactions such as

$$\begin{aligned} \pi^- + p &\to K^0 + \Lambda^0 \\ &\to K^+ + \Sigma^- \\ &\to 2K^0 + \Xi^0 \\ &\to 2K^0 + K^+ + \Omega^- \end{aligned} \tag{3-15}$$

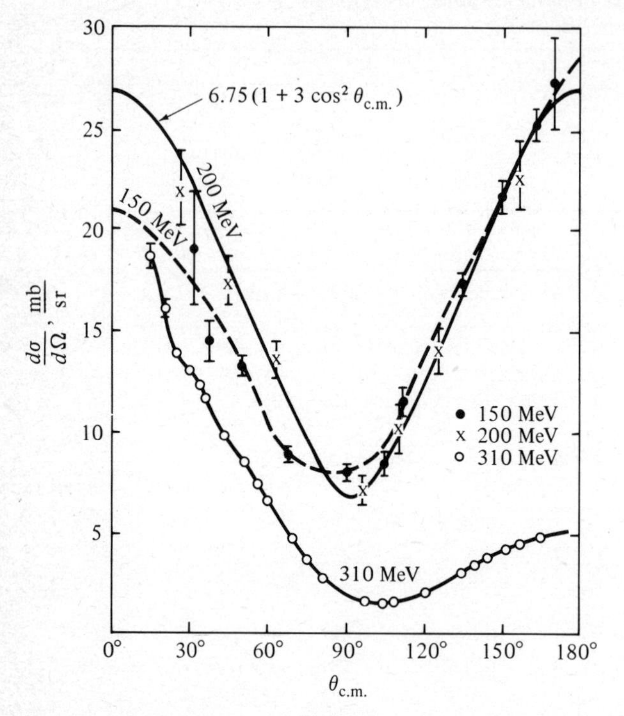

Figure 3-18 Typical π^+-p differential cross sections in the region of the $\Delta(1236)$ resonance.

These reactions have considerably lower threshold energies than reactions producing hyperons in N-N collisions (see Prob. 3-7).

3-6 OTHER SYSTEMS

The interactions of kaons and antiprotons with nucleons have been thoroughly studied over a wide range of energies. We shall not discuss these in any detail since they are generally similar to the $\pi^{\pm}$-N and N-N systems. Kaon-nucleon scattering provides a convenient means of studying resonances that have baryon number 1 and nonzero strangeness. Many of these have been found (Table

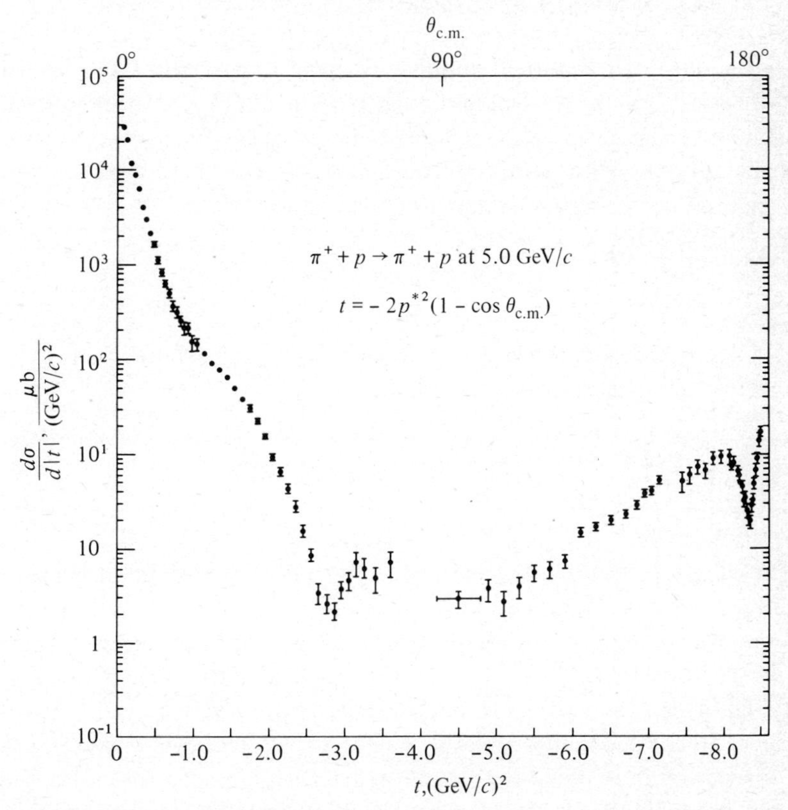

Figure 3-19 The invariant differential cross section $d\sigma/dt$ for π^+-p elastic scattering at 5 GeV/c laboratory momentum. Note that the cross section drops about five decades between 0 and 90°. *(Data provided by Professor Donald Meyer, University of Michigan.)*

3-1). An interesting feature of $\bar{p}$-p scattering is that this system has baryon number zero. This makes possible reactions such as

$$
\begin{aligned}
\bar{p} + p &\to \pi^+ + \pi^- \\
&\to \pi^+ + \pi^- + \pi^0 \\
&\to K^0 + \bar{K}^0 \\
&\to \Lambda^0 + \bar{\Lambda}^0 \\
&\underset{\text{em}}{\to} e^+ + e^- \\
&\to \cdots
\end{aligned}
$$

3-7 MESON–MESON RESONANCES

In addition to the great number of baryon resonances mentioned previously, many resonances with baryon number zero have been found. These cannot be studied directly, say in π-π scattering, but are observed as products in many-body final states. For example, in a bubble chamber one can study

$$\pi^+ + p \rightarrow \pi^+ + \pi^0 + p$$

If the π^+ and π^0 are decay products of some very short-lived state, say X^+, the actual reaction will be

$$\begin{aligned} \pi^+ + p &\rightarrow X^+ + p \\ &\rightarrow \pi^+ + \pi^0 + p \end{aligned}$$

If this is the case, the π^+ and π^0 momenta are correlated because they are decay products from a parent with a fairly well defined mass (consistent with $\Delta E\, \Delta t \approx \hbar$).

The events in the bubble chamber look roughly as shown in Fig. 3-20, where the two γ's from the π^0 decay are not observed. However if the momenta and the angles of the charged particles are measured, it is possible to reconstruct the π^0 momentum with the assumption that the reaction really is $\pi^+ + p \rightarrow \pi^+ + \pi^0 + p$. This is possible because there are three unknowns, namely, the three components of the π^0 momentum, but four constraints from conservation of energy and vector momentum.

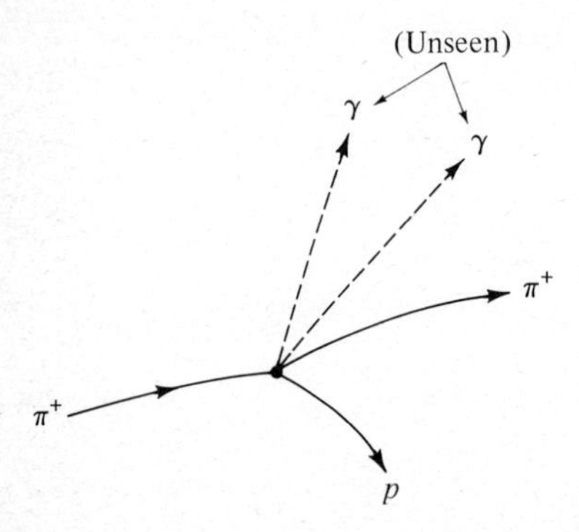

Figure 3-20 Schematic of a bubble-chamber event for the reaction $\pi^+ + p \rightarrow \pi^+ + \pi^0 + p$.

Once the momentum of the π^0 is calculated, the mass of the postulated X^+ can be determined. If the π^+ and π^0 are considered decay products of a parent particle X^+, then

$$\mathbf{p}_{\pi^+} + \mathbf{p}_{\pi^0} = \mathbf{p}_X \qquad \text{and} \qquad E_{\pi^+} + E_{\pi^0} = E_X$$

Then, if $c \equiv 1$,

$$\begin{aligned} m_X{}^2 &= E_X{}^2 - p_X{}^2 \\ &= \sum E_i{}^2 - \sum p_i{}^2 \end{aligned}$$

where, in general, the sum is over all supposed decay products of the X. The quantity m_X is often referred to as the *missing mass, effective mass,* or *invariant mass.* A resonance shows up as a bump in the distribution of m_X for the total sample of events studied. A typical plot with a well-defined peak is shown in Fig. 3-21. The peak corresponds to a resonance called the ρ^0, which decays into π^+ and π^-. The dashed line represents the sort of distribution expected if the π^+-π^- mass distribution is determined by statistical factors only [the so-called density of states (Sec. 4-1)]. This provides an "educated guess" of how the curve might look without resonances.

The lifetime of a resonance can be determined from its width and is characteristically about 10^{-24} sec for decays through strong interactions. An interesting exception to this is that of the η^0. It was discovered by using the technique described above; however the η^0 lifetime is approximately 10^{-19} sec, which is not too much shorter than that of the π^0. The η^0 therefore deserves its place on the list of more stable particles given in Chap. 1. The lifetime of the η^0 is more appropriate to a decay through electromagnetic interactions, which are, in fact, responsible for its decay. The η has spin 0 and negative parity. The decay $\eta^0 \rightarrow 2\pi$ is forbidden because two pions, being bosons, must have parity $(-1)^\ell$, which is positive when $\ell = J = 0$. The decay $\eta^0 \rightarrow 3\pi$ is also forbidden to proceed through strong interactions. This is easiest to understand by using the concept of G parity. The *G parity* is a quantum number which is conserved only in the strong interactions. Its conservation follows from a consideration of invariance under the combined operations of charge conjugation and a rotation of 180°

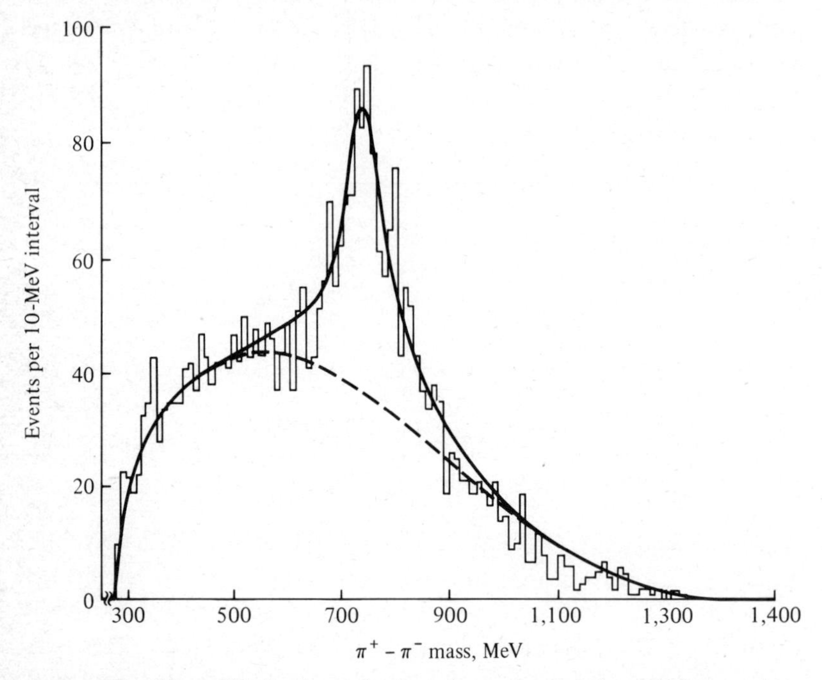

Figure 3-21 Mass distributions of π^+-π^- from the reaction $\pi^+ + p \rightarrow \pi^+ + \pi^+ + \pi^- + p$ for 2.9 GeV/c incident pions. The peak corresponds to the $\rho(765)$. The dashed curve shows the shape expected if the mass spectrum were determined by statistical factors only. [*From Alff et al., Phys. Rev. Letters* **9**, 322 (1962).]

about the Y axis in isospin space.[1] A system of n pions has G parity $(-1)^n$. The η has G parity $+1$. The decay $\eta \rightarrow 3\pi$ can therefore only take place through electromagnetic interactions. Furthermore, the decay $\eta \rightarrow 4\pi$ is barely possible energetically and is therefore greatly suppressed by kinematic factors. The most probable decay mode of the η^0 is thus $\eta \underset{\text{em}}{\longrightarrow} 3\pi$.

When higher-energy beams and large bubble chambers became

[1] See W. Frazer, "Elementary Particles," Prentice-Hall, Englewood Cliffs, N.J., 1966. Consideration of G parity provides a useful selection rule for strong interactions: If the initial state of a reaction contains n_1 pions (and no other particles), and the final state contains n_2 pions, then $n_1 + n_2$ must be an even number. G is only defined for systems with baryon number zero and zero strangeness. The G-parity operation is the analog in isotopic spin space of the parity operation in ordinary space.

available, the elementary particle "population" underwent an explosion as more and more short-lived particles were found. There are now approximately 100 known, and the list increases almost monthly. There are K-π resonances, K-K resonances, and even resonances which decay into other resonances, e.g.,

$$B(1235) \rightarrow \omega(784) + \pi$$

The quantum numbers of the higher mass resonances are often difficult to establish. The isospin can usually be determined from the charge states found. For example, if a meson state appears as X^+, X^0, and X^-, but not as X^{++} or X^{--}, it has an isospin of 1 like the pion. The spin can often be established from the angular distribution of the decay products. Most of the resonances can decay in a variety of ways. A list of *some* of the better established meson and baryon resonances is given in Table 3-1.

PROBLEMS

3-1 What fraction of an incident 2-GeV π^+ beam will interact in a 1-m-long liquid-hydrogen target? ($\rho = 0.07$ g/cm^3.)

3-2 If a beam of 10^5 pions per second is incident on a liquid-hydrogen target 10 cm long, how many elastically scattered pions will be detected per second in a counter 10 cm square, 1 m away from the target? Take $(d\sigma/d\Omega)_{\text{lab}} = 10$ mb/sr at the angle of the detector.

3-3 What is the de Broglie wavelength for a 5-GeV pion? Calculate the angle of the first diffraction minimum for light of this wavelength and a circular aperture of 1-fm radius.

3-4 What predictions about the following reactions can be made from isotopic spin considerations?

(*a*) $d + d \rightarrow d + d + \pi^0$

(*b*) $d + d \rightarrow \text{He}^4 + \pi^0$ (He^4 has isospin 0)

3-5 A model for the technique of dynamic polarization that is used to polarize protons consists of a system of one proton and one electron in a magnetic field **B**. There are four possible spin states of the system $(\frac{1}{2},\frac{1}{2})$, $(\frac{1}{2},-\frac{1}{2})$, $(-\frac{1}{2},\frac{1}{2})$, and $(-\frac{1}{2},-\frac{1}{2})$. The energy levels are as shown in Fig. 3-22, where Δ is about $10^3\delta$.

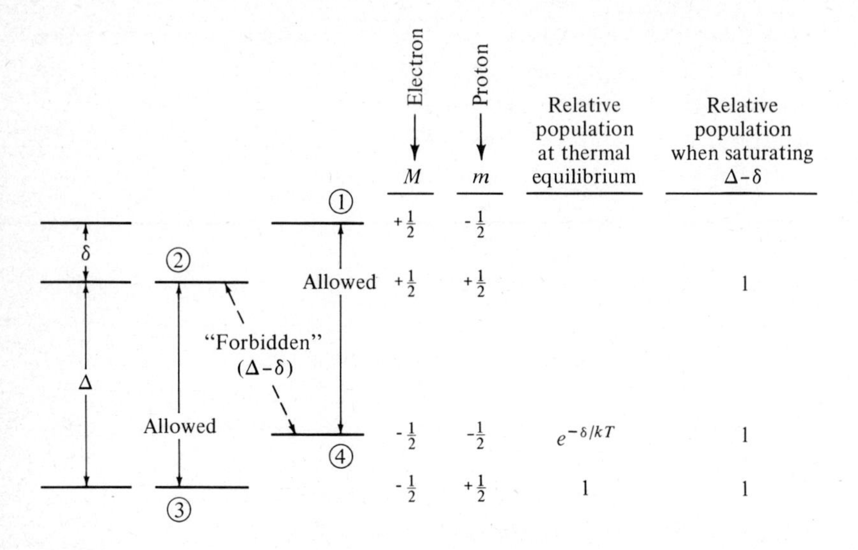

Figure 3-22 Energy-level diagram for a system composed of one proton and one neutron (level spacings are not drawn to scale).

(*a*) Complete the column showing the relative population of the levels at thermal equilibrium. Calculate the electron polarization $(N_+ - N_-)/(N_+ + N_-)$ if $T = 1°\text{K}$ and $B = 20$ kG. Show that the proton polarization is quite small under these conditions ($\approx 1 + \delta/kT$).

(*b*) A strong microwave generator with a frequency appropriate for inducing the transition $\Delta - \delta$ is now turned on. If the microwaves are strong enough to saturate the transition, the population of levels 2 and 4 equalize, with level 3 in thermal equilibrium with level 2, and 1 in equilibrium with 4. Complete the column showing the relative populations under these conditions. Calculate the proton polarization, and show that it is approximately equal to the initial electron polarization.

(*c*) How can the sign of the proton polarization be changed without reversing the magnetic field?

3-6 Calculate the angular distribution for the scattering of 1-GeV/c spinless particles if there is complete absorption for $\ell \leq 3$ and no effect (zero phase shift) on partial waves with $\ell > 3$. Make a plot of the angular distribution. By graphical integration, calculate the total elastic cross section. Compare this with σ_T from the optical theorem and with $\sigma_{el} = \pi\lambda\!\!\!^{-2} \sum (2\ell + 1)|1 - \eta_\ell|^2$. Explain any discrepancies.

3-7 What is the reaction with the lowest threshold for producing a Λ^0 in a π^--p collision (proton at rest)? In a p-p collision? Compare the threshold energies for the two reactions. What is the reaction with the lowest threshold for producing $\bar{\Lambda}^0$ in a π^--p collision?

3-8 What is the minimum pion-beam energy for producing Ξ^0's in π^--p collisions through strong interactions?

3-9 Design an experiment to measure total cross sections for π^+ on protons in the momentum range 1 to 3 GeV/c with an accuracy of ± 1 percent. Consider the following:
(*a*) Choice of an accelerator
(*b*) General features of beam design
(*c*) The technique and apparatus
(*d*) Statistical and systematic errors

3-10 Below 200 MeV, experiments show that $f_{\frac{1}{2}}$ for π-N scattering is very small.
(*a*) Use this result together with the data in Fig. 3-15 to predict the total cross section for $\pi^- + p \rightarrow \pi^0 + n$ below 200 MeV. (At these energies absorption can be neglected.)
(*b*) Use this result to "predict" $\sigma_T(\pi^-$-$p)$ from $\sigma_T(\pi^+$-$p)$, and compare the result with the experimental results.

3-11 Estimate ℓ_{max} from Eq. (3-8) for the scattering of pions with a laboratory momentum of 1 GeV/c on protons.

3-12 Estimate the probability of a 2-GeV pion interacting in traversing a carbon nucleus if the total cross section for bound nucleons equals that for free nucleons. (The nuclear radius is $\approx 1.3A^{\frac{1}{3}} \times 10^{-13}$ cm.) Repeat for a lead nucleus.

3-13 Use the optical theorem to calculate $d\sigma/dt$ at $t = 0$ for π^+-p scattering at 5 GeV/c. [For elastic scattering,

$$t = -2p^2(1 - \cos\theta_{c.m.})$$

where p is the momentum in the c.m.s. For π-p scattering, 5 GeV/c in the laboratory corresponds to 1.46 GeV/c in the c.m.s.] Compare this with the experimental value from Fig. 3-19.

3-14 Use Eq. (3-14) to verify the spin assignment for the $\Delta(1236)$, assuming the cross section reaches the maximum value for $J = \frac{3}{2}$ at the peak of the resonance.

3-15 In the Born approximation the scattering amplitude for scattering from a spherically symmetrical potential well $U(r)$ is

$$f(\theta) = \frac{-2m}{\hbar q} \int rU(r) \sin \frac{qr}{\hbar} \, dr$$

where $q = 2p \sin(\theta/2)$.[1] Use this to calculate the differential cross section for scattering from a square well ($U = -U_0$ for $r < r_0$; $U = 0$ for $r > r_0$). Make a rough plot for $U_0 = 25$ MeV, $r_0 = 2$ fm, $p = 2$ GeV/c.

3-16 Use G parity (Sec. 3-7) to show that diagrams like that shown here are not allowed in π-p elastic scattering.

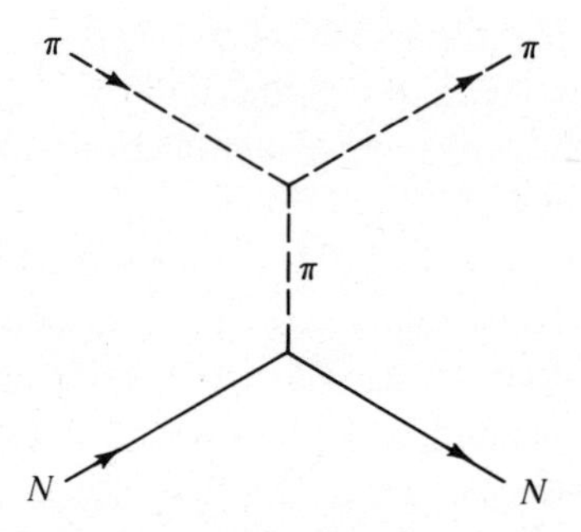

What is the least massive particle with the correct quantum numbers (parity, G parity, etc.) in Tables 1-1 and 3-1 that can be exchanged in the diagram? What implication does this have for the range of the π-N interaction compared with that of the N-N interaction?

3-17 Use the Bohr theory of the hydrogen atom to estimate the magnetic field at the nucleus due to the orbital motion of the electron in the ground state.

[1] Schiff, *op. cit.*

CHAPTER 4

calculation of rates for various kinds of processes

4-1 INTRODUCTION

We shall now discuss in a very qualitative way how rates for various kinds of processes such as decays can be estimated. Let us compare the following decays:

1.	$N^*(1236) \to p + \pi^+$	Strong, $\tau \sim 6 \times 10^{-24}$ sec
2.	$\Sigma^0 \to \Lambda^0 + \gamma$	Electromagnetic, $\tau \lesssim 10^{-14}$ sec
3.	$\Sigma^+ \to n + \pi^+$	Weak, $\tau \sim 10^{-10}$ sec

These processes are similar in many respects. The main difference is that the first proceeds through a strong interaction, whereas the second involves a γ and therefore goes by an electromagnetic interaction, and the third violates strangeness and must therefore go by a weak interaction.

4-2 GOLDEN RULE NUMBER 2

We start with a well-known formula from quantum mechanics which is sometimes referred to as *Golden Rule Number* 2.[1] The transition rate for a reaction $a \to b$ is expressed in terms of a *matrix element* M_{ba} as

$$(\text{Trans. rate})_{a\to b} \propto |M_{ba}|^2 \rho(E_i) \tag{4-1}$$

For a decay, the transition rate is equal to the reciprocal of the lifetime τ, and for a scattering process, it is proportional to the

[1] See, for example, L. I. Schiff, "Quantum Mechanics," 3d ed., chaps. 8 and 9, McGraw-Hill, New York, 1968.

scattering cross section. The quantity $\rho(E_i)$ is a kinematical factor called the *density of states,* which gives the number of available energy states per unit energy range. For example, in a two-body final state in which one particle is much more massive than the other,

$$\rho(E) \propto E_1 p_1$$

where E_1 and p_1 are the total energy and momentum of the lighter particle. Since $\rho(E_i)$ is roughly comparable for all three of the reactions above, it can be disregarded in our crude comparison.

4-3 ORDER-OF-MAGNITUDE ESTIMATES OF MATRIX ELEMENTS

The density of states can be calculated in a straightforward way for all reactions. The real physics comes in calculating the matrix element M_{ba}. Physically M_{ba} is a measure of the intrinsic probability of state a's going to state b. In principle it can be calculated from the complete wave functions of these states and the hamiltonian of the interaction, but in practice for interactions including *hadrons* it is very difficult to make more than a crude estimate of M_{ba}.

For our purpose it is sufficient to note that in general $|M_{ba}|^2 \sim g^2$, where g is the coupling constant appropriate to the type of interaction involved in the process $a \to b$. For strong, electromagnetic, and weak reactions the order of magnitude of the coupling constant is as follows:[1]

For strong interactions: $g_s^2 \sim 1$

For electromagnetic interactions: $g_{\text{em}}^2 = \dfrac{e^2}{\hbar c} \approx \frac{1}{137}$

For weak interactions: $g_w^2 \sim 10^{-14}$

On this basis the ratio of the decay rates for the three decays considered above should be $1:6 \times 10^{-3}:10^{-14}$, which agrees reasonably well with the ratios of observed rates (the decay rate $= \tau^{-1}$).

[1] This is a rather crude specification of the coupling constants, but it is adequate for our purpose.

Richard P. Feynman. Born 1918 in New York City. He received his Ph.D. in 1942 from Princeton. At Princeton he became interested in the problems of emission and absorption of radiation by charged particles. During the war he worked on the atomic bomb project. After the war while at Cornell University, he developed his elegant graphic representation

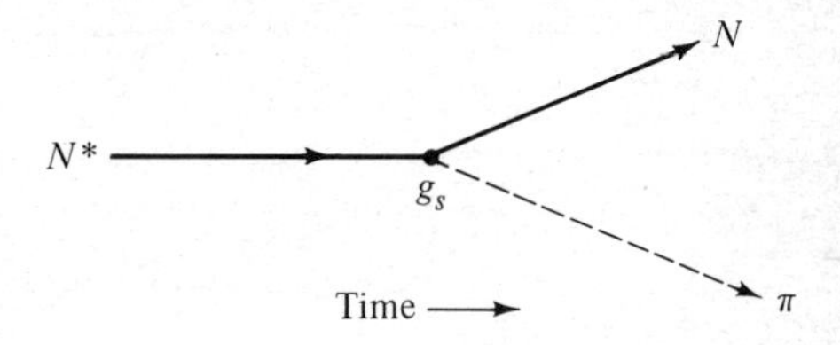

Figure 4-1 Decay of $N^*(1236)$ [or $\Delta(1236)$] into a nucleon and pion.

In generalizing this procedure to other processes, it is convenient to use the language of *Feynman diagrams.* For the N^* decay one such diagram is shown in Fig. 4-1. The rule is that for every vertex in a Feynman diagram, the matrix element contains a factor of g, where g is the coupling constant appropriate to the type of reaction which occurs at that vertex. Thus for N^* decay, M_{ba} contains a factor g_s, and the decay rate, which is proportional to $|M_{ba}|^2$, is therefore proportional to g_s^2. If a diagram contains two vertices, the matrix element M_{ba} includes a product of two coupling constants. Thus in the diagram in Fig. 4-2, which contains two strong vertices, the matrix element is proportional to g_s^2. A diagram like this with two vertices is called a *second-order* diagram. The difficulty in making calculations for processes involving hadrons now becomes apparent. The overall decay rate for a process is obtained by somehow summing the contributions from all possible diagrams for the initial and final states under consideration. Since $g_s \sim 1$, second- and higher-order strong processes give contributions comparable to the first-order ones. The situation is much more favorable for electromagnetic and weak processes. In this case higher-order processes are suppressed by the additional factors of g_{em} or g_w, which are $\ll 1$.

To represent the electromagnetic process of Σ^0 decay, we cannot draw a diagram like that in Fig. 4-3a since uncharged particles do not emit γ's. We can easily get around this by invoking the strong

---

for quantum electrodynamic processes. In a series of historic papers Feynman laid much of the groundwork for modern quantum electrodynamics. In 1965 he was corecipient of the Nobel Prize with Julian Schwinger and Schinichiro Tomonaga for their work in the development of quantum electrodynamics. In recent years he has turned his attention to elementary particles and low-temperature physics. He is the author of several advanced books on theoretical physics and in addition is a popular lecturer on undergraduate physics. (*Photograph from the Marshak Collection of the Niels Bohr Library, American Institute of Physics.*)

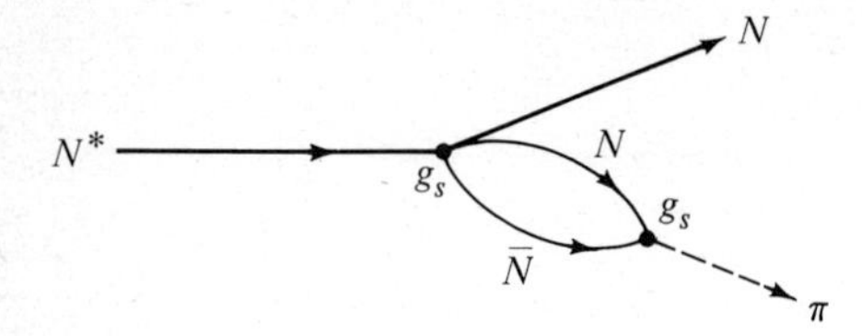

Figure 4-2 Second-order diagram for decay of $N^*(1236)$.

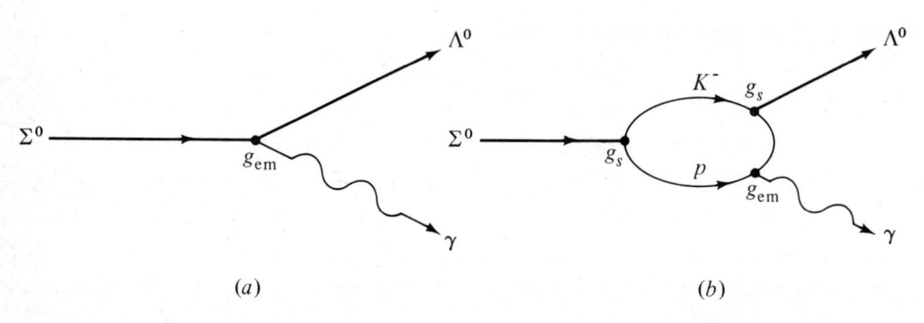

Figure 4-3 (*a*) Unallowed diagram for decay $\Sigma^0 \rightarrow \Lambda^0 + \gamma$. (*b*) Possible diagram which makes use of virtual dissociation of Λ^0 into $K^- + p$ through strong interactions.

interactions as shown in Fig. 4-3*b*. The matrix element for Fig. 4-3*b* is proportional to $g_s{}^2 \cdot g_{\text{em}}$. Since $g_s \sim 1$, the matrix element is effectively $\sim g_{\text{em}}$. There are obviously many other possible diagrams for this process, but all have two or more factors of g_s and at least one factor of g_{em} so that $M_{ba} \sim g_{\text{em}}$.

As an example of a second-order electromagnetic process, consider Compton scattering (Sec. 2-3), the simplest diagram for which is shown in Fig. 4-4. Since there are two electromagnetic vertices, the matrix element is proportional to $(g_{\text{em}})^2 = \frac{1}{137}$. This process

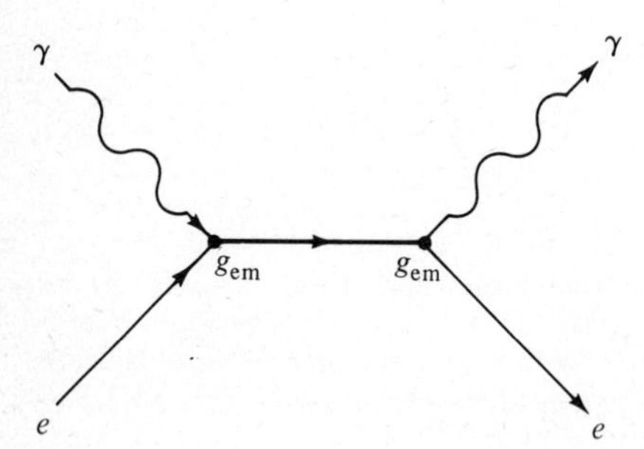

Figure 4-4 Compton scattering of a γ from an electron, an example of a second-order electromagnetic process.

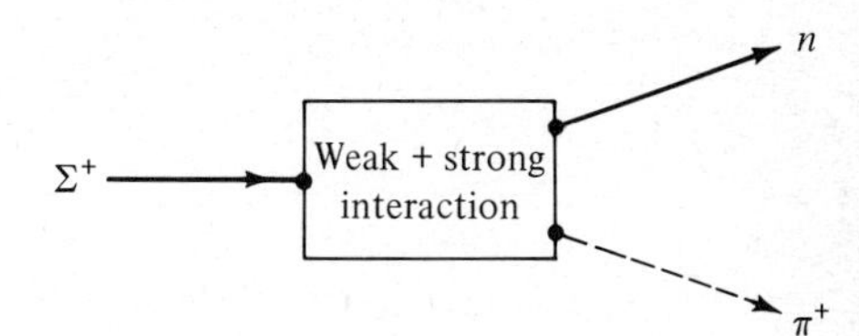

Figure 4-5 Weak decay of Σ^+ to $n + \pi^+$.

involves no hadrons and therefore can be calculated to a high degree of accuracy.

The diagram for the weak decay of the Σ^+ is shown in Fig. 4-5. We do not know exactly what is in the "box," but the rate will be controlled by g_w and the matrix element will be proportional to g_w. (We shall discuss the contents of the box further in Chap. 6.)

4-4 MORE ON FEYNMAN DIAGRAMS. APPLICATION TO SCATTERING PROCESSES

The rules for drawing Feynman diagrams are rather simple. At a vertex we must always conserve charge, baryon number, and lepton number. At a strong vertex we must also conserve strangeness (and G parity where appropriate). Strangeness must also be conserved at an electromagnetic vertex, but it need not be conserved at a weak vertex. Energy and momentum must of course be conserved in the overall process and can be formally considered conserved at each vertex. However, the relation $m^2c^4 = E^2 - p^2c^2$ is not valid for intermediate (virtual) particles. The diagrams also have the property that the direction of any line can be reversed if the corresponding particle is changed to its antiparticle. Thus the diagram for π-p elastic scattering in Fig. 4-6a can be transformed to that for $\bar{p} + p \rightarrow \pi^\pm + \pi^\mp$ by changing the incident $\pi^\pm$ to an outgoing $\pi^\mp$ and the outgoing p to an incoming $\bar{p}$.

We can draw diagrams for scattering processes also (as we have already done in Chap. 3) and use them to estimate cross sections. For example, by comparing the Feynman diagram for $\pi + p \rightarrow \pi + p$ with that for the process $\gamma + p \rightarrow \pi^0 + p$ (Fig. 4-6b), we see that the latter contains an additional factor of g_{em}. Thus we estimate

$$\sigma_T(\gamma + p \rightarrow \pi^0 + p) \sim \tfrac{1}{137}\sigma_T(\pi + p \rightarrow \pi + p) \qquad (4\text{-}2)$$

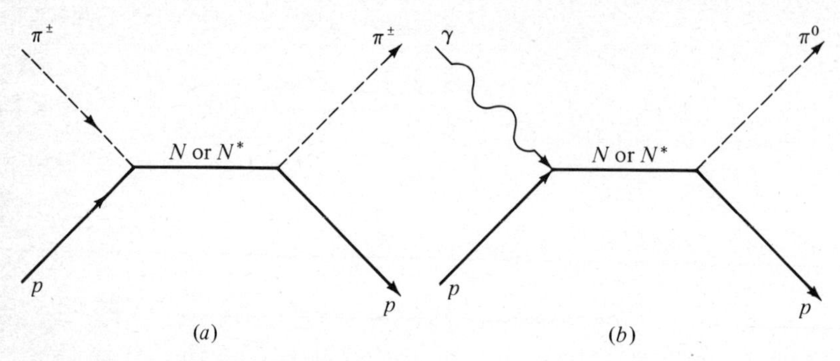

Figure 4-6 (*a*) Diagram for $\pi^{\pm} + p \rightarrow \pi^{\pm} + p$. (*b*) Diagram for $\gamma + p \rightarrow \pi^{0} + p$.

At high energies $\sigma_T(\pi + p \rightarrow \pi + p) \approx 30$ mb, so from Eq. (4-2) we estimate $\sigma_T(\gamma + p \rightarrow \pi^0 + p) \sim 0.2$ mb. Since the experimental value of the latter cross section is ≈ 0.04 mb, our estimate is not unreasonable. This illustrates an important point. There are often factors such as angular-momentum or isotopic spin effects that may suppress (or, somewhat less likely, enhance) reaction rates and cause estimates based only on coupling constants to be in error.

The above discussion, though crude and oversimplified, may help to give the uninitiated reader some feeling for the factors which dominate the calculation of a rate or cross section. The use of diagrams for such calculations developed from their original application to quantum electrodynamics, where they are useful in identifying the terms in a perturbation series expansion. Their use in connection with strong interactions may however be misleading since perturbation theory has in fact met with very little success when applied to strong interactions.

PROBLEMS

4-1 Estimate the ratio of rates for the following reactions, and compare with the experimental results (Table 1-1).

(*a*) $(\pi^0 \rightarrow \gamma + e^+ + e^-)/(\pi^0 \rightarrow \gamma + \gamma)$

(*b*) $(\Sigma^+ \rightarrow p + \gamma)/(\Sigma^+ \rightarrow p + \pi^0)$

(*c*) $(K^0 \rightarrow 3\pi)/(\eta^0 \rightarrow 3\pi)$

4-2 What are the simplest diagrams for the following reactions, consistent with the conservation laws?

(*a*) π^--p elastic scattering
(*b*) K^--p elastic scattering
(*c*) e-p elastic scattering
(*d*) $\pi^- + p \rightarrow K^+ + \Sigma^-$
(*e*) $\pi^- + p \rightarrow K^0 + \Lambda^0$
(*f*) γ-p elastic scattering
(*g*) $\gamma + p \rightarrow \pi^+ + n$
(*h*) $\bar{p} + p \rightarrow \pi^+ + \pi^-$
(*i*) $\bar{p} + p \rightarrow e^+ + e^-$
(*j*) $\pi^0 \rightarrow \gamma + \gamma$

4-3 Devise two possible diagrams other than that shown in Fig. 4-3*b* for the decay $\Sigma^0 \rightarrow \Lambda^0 + \gamma$.

CHAPTER 5

electromagnetic interactions

Electromagnetic interactions constitute a rich and varied topic; we shall discuss only the most significant features of their high-energy behavior. We neglect entirely the low-energy domain, of which atomic and molecular physics are notable examples. The theoretical treatment of electromagnetic interactions is the subject of quantum electrodynamics. Unfortunately the techniques used in calculations are too advanced to be given here. We shall therefore only quote the results and their application to the experimental data.

5-1 ELECTRON-SCATTERING EXPERIMENTS

Experiments to study electron scattering from nucleons and nuclei have provided a great deal of information on electric-charge distributions. The first-order Feynman diagram for elastic *e-p* scattering is shown in Fig. 5-1. The box at the nucleon vertex represents

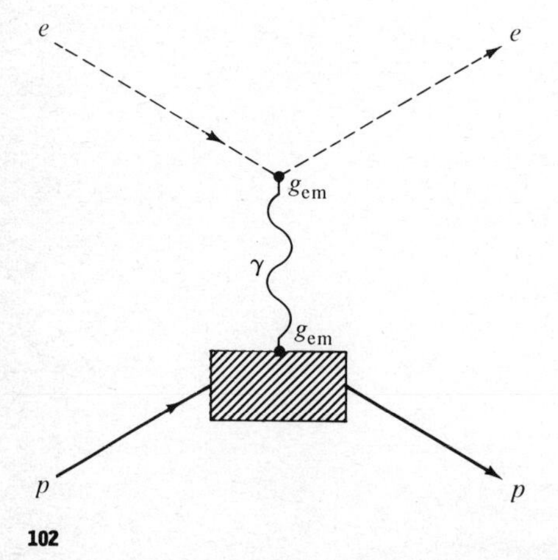

Figure 5-1 Diagram for *e-p* elastic scattering. The box represents the effects of virtual strong interactions.

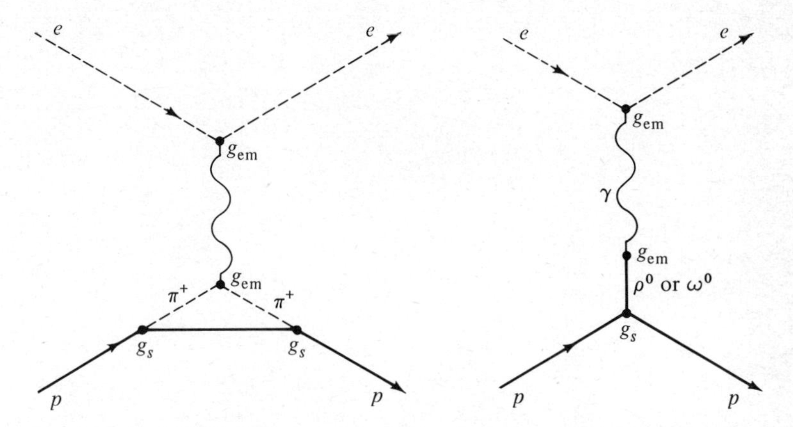

Figure 5-2 Diagrams for *e-p* elastic scattering, showing possible strong processes at lower vertex.

schematically the effects of virtual strong interactions which effectively surround the proton with a cloud of hadrons. For example, processes like those in Fig. 5-2 can occur in the box.

If it were not for the virtual strong interactions, the differential cross section for *e-N* scattering could be calculated exactly by the methods of quantum electrodynamics. In practice we turn the problem around and use the measured differential cross sections to learn about the charge distribution of the nucleon, which is determined by the strong interactions. The technique is completely analogous to that used by Rutherford and others for determining the charge distribution in the atom by studying the scattering of alpha particles. The main difference is the probe (electrons instead of alphas) and the size of the target ($\sim 10^{-13}$ cm instead of $\sim 10^{-8}$ cm). To resolve such a small distance, the wavelength of the incident electrons, h/p, must be $\lesssim 10^{-13}$ cm. This requires that p be $\gtrsim 1.2$ GeV/c.

Most of the early electron-scattering work was done by R. Hofstadter and his collaborators at Stanford University. This early work has been extended to higher energies with the 20-GeV electron accelerator at SLAC and those at several other laboratories (see Chap. 2). The technique is shown schematically in Fig. 5-3. The momentum and angle of either the elastically scattered electron or the recoil proton (or both) is measured. Figure 5-4 shows the

Robert L. Hofstadter. Born 1915 in New York. He received his Ph.D. from Princeton in 1938. In 1948 Hofstadter showed that sodium iodide, activated by thallium, made an excellent scintillation counter. In the 1950s at Stanford University he and his colleagues carried out a series of classic experiments to investigate the scattering of electrons from nucleons and nuclei. These led to a rather detailed knowledge of the charge and magnetic-moment distributions of protons, neutrons, and nuclei. In 1961 he shared the Nobel Prize with Rudolf Mössbauer. (*Photograph from the Meggers Gallery of Nobel Laureates, American Institute of Physics.*)

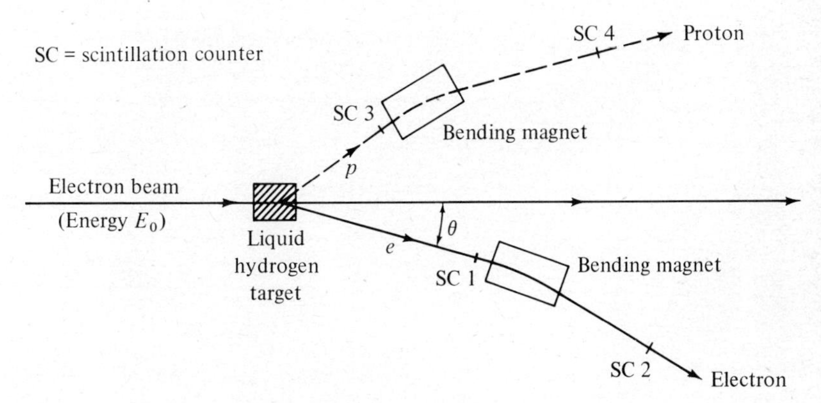

Figure 5-3 Schematic of an *e-p* scattering experiment. Either or both of the final-state particles are detected.

huge spectrometers used at SLAC for this purpose. Electron-neutron scattering can be studied by using a deuterium target and making appropriate corrections.

The differential cross section for elastic electron scattering from a point charge at an angle θ in the laboratory is

$$\frac{d\sigma}{d\Omega} \equiv \sigma_{\text{point}}(\theta) = \frac{Z^2\alpha^2 \cos^2(\theta/2)}{4E_0{}^2 \sin^4(\theta/2)} \frac{1}{1 + 2(E_0/M)\sin^2(\theta/2)} \tag{5-1}$$

where $\alpha \approx \frac{1}{137}$, E_0 is the electron energy in the laboratory, M is the mass of the target nucleon or nucleus, Z is its atomic number, and we choose units such that $\hbar = c = 1$.

The generalization of this formula to an extended charge distribution is[1]

$$\sigma = \sigma_{\text{point}}|F(q^2)|^2 \qquad \text{no magnetic moment} \tag{5-2}$$

where F is the so-called *charge form factor*, and q^2 is the square of the invariant four-momentum transfer in the scattering.[2] For a point

[1] See R. Hofstadter, *Ann. Rev. Nucl. Sci.* **7**, 231 (1957).

[2] $q^2 \equiv |t|$, where t is defined in Sec. 3-2. See Appendix B for a summary of relativistic formulas.

Figure 5-4 The 8 GeV/c (foreground) and 20 GeV/c (background) magnetic spectrometers at SLAC. Electrons, scattered from a target off the picture to the left, are momentum-analyzed and focused on detectors housed in the large concrete shields at the right. The spectrometers are mounted on rails to facilitate changing the scattering angle.

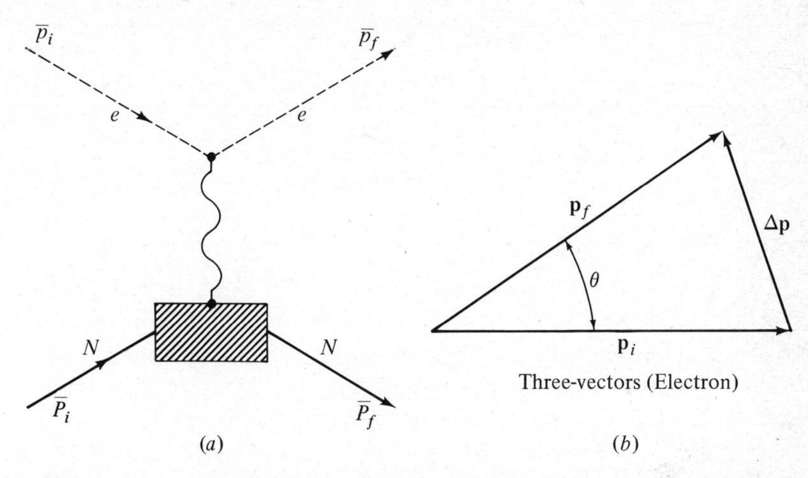

Figure 5-5 (*a*) Diagram for *e-N* elastic scattering, showing how four-vectors are defined. (*b*) Momentum three-vectors for the electron in the laboratory system. θ is the scattering angle, and $\mathbf{\Delta p}$ the momentum transfer in the collision.

charge, obviously $F = 1$. The requirement that the theory be Lorentz invariant leads to the condition that F be a function only of q^2. The momentum four-vectors are defined in Fig. 5-5*a*, where we use a bar over the symbol to denote a four-vector. The components of the four-vectors are

$$\bar{p}_i \equiv (\mathbf{p}_i, E_i^e) \qquad \bar{p}_f \equiv (\mathbf{p}_f, E_f^e)$$
$$\bar{P}_i \equiv (\mathbf{P}_i, E_i^N) \qquad \bar{P}_f \equiv (\mathbf{P}_f, E_f^N)$$

and

$$\begin{aligned} \bar{q} &\equiv (\mathbf{p}_f - \mathbf{p}_i,\ E_f^e - E_i^e) \\ &= (\mathbf{P}_i - \mathbf{P}_f,\ E_i^N - E_f^N) \end{aligned}$$

The momentum three-vectors are shown in Fig. 5-5*b*. If the target nucleon is at rest, $|\mathbf{P}_i| = 0$ and $E_i^N = M$, the mass of the nucleon. A simple expression for q^2 in terms of the kinetic energy of the recoil nucleon T can then be developed as follows:

$$\begin{aligned} q^2 &\equiv (\bar{P}_i - \bar{P}_f) \cdot (\bar{P}_i - \bar{P}_f) \\ &= \bar{P}_i^2 + \bar{P}_f^2 - 2\bar{P}_i \cdot \bar{P}_f \\ &= -2M^2 + 2(E_i^N E_f^N - \mathbf{P}_i \cdot \mathbf{P}_f) \\ &= -2M^2 + 2M(M + T) \\ &= 2MT \end{aligned}$$

Thus $q^2 = 2MT$, where T is the kinetic energy of the recoil nucleon (or nucleus).

We can also express q^2 in terms of θ and $E_0 = E_i^e$,

$$q^2 = \frac{4E_0^2 \sin^2(\theta/2)}{1 + 2(E_0/M)\sin^2(\theta/2)} \tag{5-3}$$

If $M \gg E_0$, then $q \approx 2E_0 \sin(\theta/2)$.

In the Born approximation (Prob. 3-15), $F(q^2)$ is the Fourier transform of the charge distribution in the nucleus.[1] For a spherically symmetrical nucleus with charge density $\rho(r)$, normalized so that $\int \rho(r)\,\mathbf{dr} = 1$, in this approximation

$$\begin{aligned} F(q^2) &= \frac{4\pi}{q}\int_0^\infty \rho(r)\sin(qr)r\,dr \\ &\approx 1 - \frac{q^2}{3!}\int r^2\rho(r)\,\mathbf{dr} + \frac{q^4}{5!}\int r^4\rho(r)\,\mathbf{dr} - \cdots \end{aligned} \tag{5-4}$$

where we have used the series expansion for sin qr. The form factor $F(q^2)$ can thus be expressed in terms of the moments of the charge distribution.

Unfortunately the relation between the form factor and charge distribution given in Eq. (5-4) holds only in the nonrelativistic Born approximation, and no general relation has been developed. Equation (5-4) is an aid in visualizing the charge distribution, but it is more in keeping with current ideas to work directly with the form factors.

There is an additional complication in the case of scattering from nucleons, due to the presence of the nucleon magnetic moment. As a result it is necessary to include two form factors in Eq. (5-2).[1] One of these, $G_E(q^2)$, can be thought of as describing the distribution of charge; the other, $G_M(q^2)$, the distribution of the current, or magnetic moment. The differential cross section is then[2]

$$\sigma = \sigma_{\text{point}}\left[\frac{G_E^2 + (q^2/4M^2)\mu^2 G_M^2}{1 + q^2/4M^2} + \frac{q^2}{2M^2}\mu^2 G_M^2 \tan^2\frac{\theta}{2}\right] \tag{5-5}$$

[1] Hofstadter, *loc. cit.*

[2] Equation (5-5) holds for both proton and neutron targets with the substitution of the appropriate values of G_E, G_M, and μ; σ_{point} is the same for either. In the literature, various linear combinations of G_E and G_M are often used. Ours are defined in such a way that $G_E{}^p = G_M{}^p = G_M{}^n = 1$ and $G_E{}^n = 0$ for $q^2 = 0$.

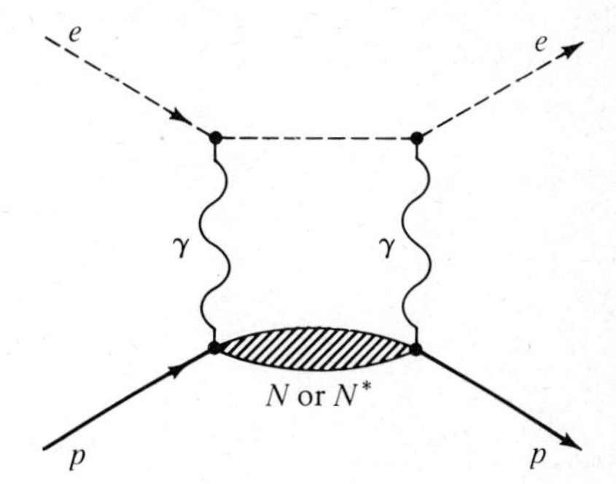

Figure 5-6 Two-photon-exchange diagram for *e-p* elastic scattering.

where μ is the nucleon magnetic moment in units $e\hbar/2M$ (for protons $\mu_p = 2.793$, and for neutrons $\mu_n = -1.91$). Equation (5-5) is often referred to as the *Rosenbluth formula.*

The Rosenbluth formula should be correct provided contributions to the cross sections from two-photon-exchange diagrams (Fig. 5-6) are negligible. Such contributions are expected to be of relative order $\frac{1}{137}$,[1] but could become more important at large q^2. The Rosenbluth formula requires that

$$\frac{\sigma}{\sigma_{\text{point}}} = a(q^2) + b(q^2) \tan^2 \frac{\theta}{2}$$

It can therefore be tested by plotting data for a given q^2 versus $\tan^2 (\theta/2)$; a deviation from a straight line would be an indication of a failure of the one-photon-exchange approximation. Another test is to compare the scattering of positrons with that of electrons. If only one-photon exchange is important, Eq. (5-5) describes positron scattering as well as electron scattering. Two-photon-exchange contributions to positron scattering, however, are opposite in sign to those for electron scattering; this would cause e^+-p cross sections to differ from the e^--p. All the experimental results to date are consistent with pure one-photon exchange.

[1] The discerning reader may wonder why this factor should not be $\frac{1}{137}{}^2$, since the two-photon-exchange diagram in Fig. 5-6 contains two more electromagnetic vertices than the one-photon-exchange diagram (Fig. 5-1) has. The factor $\frac{1}{137}$ arises because in calculating the cross section the amplitudes for one-photon exchange and two-photon exchange are added and then squared. In addition to the two-photon-exchange amplitude squared (of relative order $\frac{1}{137}{}^2$), there is a cross term, or *interference term,* of relative order $\frac{1}{137}$.

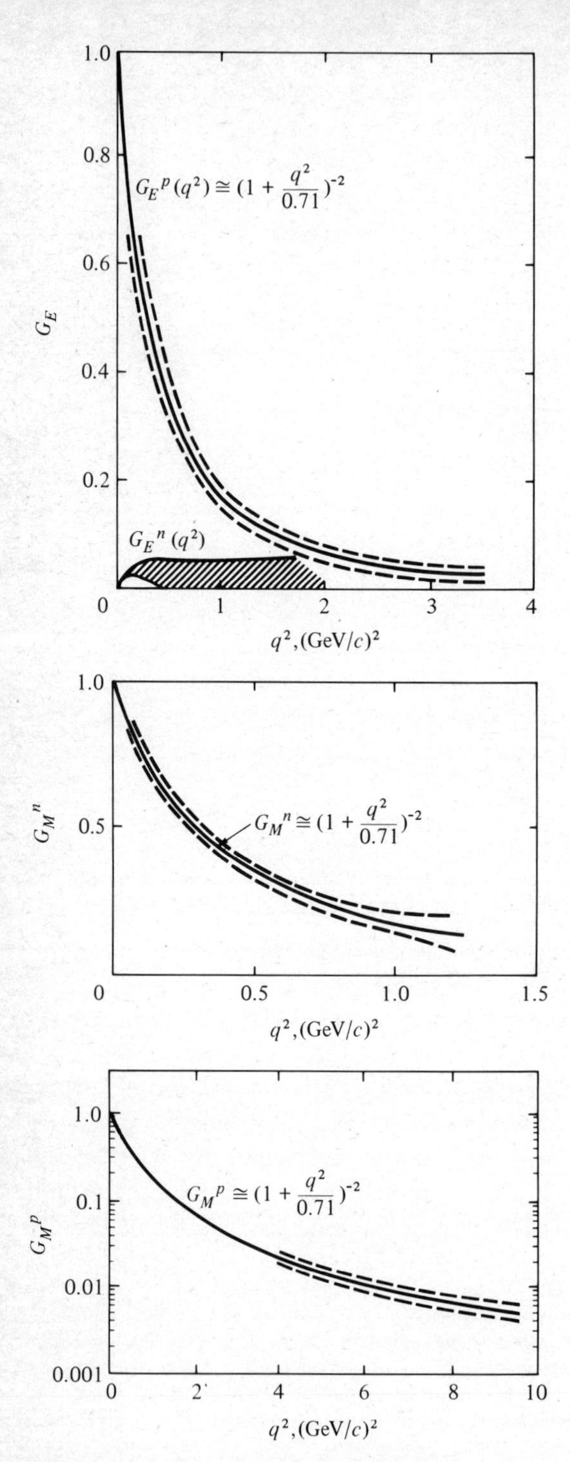

$G_E^p(q^2) \cong (1 + \frac{q^2}{0.71})^{-2}$
$G_E^n(q^2)$
G_E
1.0
0.8
0.6
0.4
0.2
0
1
2
3
4
q^2, (GeV/c)^2
$G_M^n \cong (1 + \frac{q^2}{0.71})^{-2}$
G_M^n
1.0
0.5
0
0.5
1.0
1.5
q^2, (GeV/c)^2
$G_M^p \cong (1 + \frac{q^2}{0.71})^{-2}$
G_M^p
1.0
0.1
0.01
0.001
0
2
4
6
8
10
q^2, (GeV/c)^2

5-2 DISCUSSION OF FORM FACTORS

The form factors G_E and G_M can be determined by measuring $\sigma/\sigma_{\text{point}}$ over a large range of E_0 and q^2. The results for the neutron and proton are shown in Fig. 5-7. To a good approximation the data can be fitted by the empirical formulas

$$G_E{}^p(q^2) \approx G_M{}^p(q^2) \approx G_M{}^n(q^2) \approx \left(1 + \frac{q^2}{0.71(\text{GeV}/c)^2}\right)^{-2} \tag{5-6}$$

We also see from Fig. 5-7 that $G_E{}^n(q^2) \approx 0$. A *qualitative* picture of the nucleon charge distributions can be obtained by taking the Fourier transform of the charge form factors. The results are not unique since the data can be reproduced with various distributions. One possibility is shown in Fig. 5-8. The root-mean-square (rms) charge radius of the proton is found to be approximately 0.8 fm. For comparison, the charge distributions of several nuclei are shown in Fig. 5-9. Nuclear radii are found to be given by

$$r \approx 1.3A^{\frac{1}{3}} \qquad \text{fm} \tag{5-7}$$

Note that at the largest q^2 studied, the form factors for the proton are $\sim 10^{-2}$. The cross sections at large q^2 are therefore $\sim 10^{-4}$ times those expected from a point charge. The proton might thus be described as rather "mushy" with no pointlike "hard" core.

Some progress has been made in interpreting the experimental results for nucleon form factors within our present understanding of strong interaction dynamics. Good fits to the data have been obtained by assuming that the scattering is dominated by vector meson (spin 1^-) exchange diagrams of the type shown in Fig. 5-10.[1]

[1] E. B. Hughes, T. A. Griffy, M. R. Yearian, and R. Hofstadter, *Phys. Rev.* **139,** B458 (1965).

Figure 5-7 Experimental results for the proton and neutron form factors. The dashed lines and crosshatched bands indicate the experimental uncertainty. For $q^2 > 0.2\ (\text{GeV}/c)^2$, the form factor $G_E{}^n$ is consistent with zero. All the other form factors are well described by the empirical formula $(1 + q^2/0.71)^{-2}$. (Note log scale for $G_M{}^p$.)

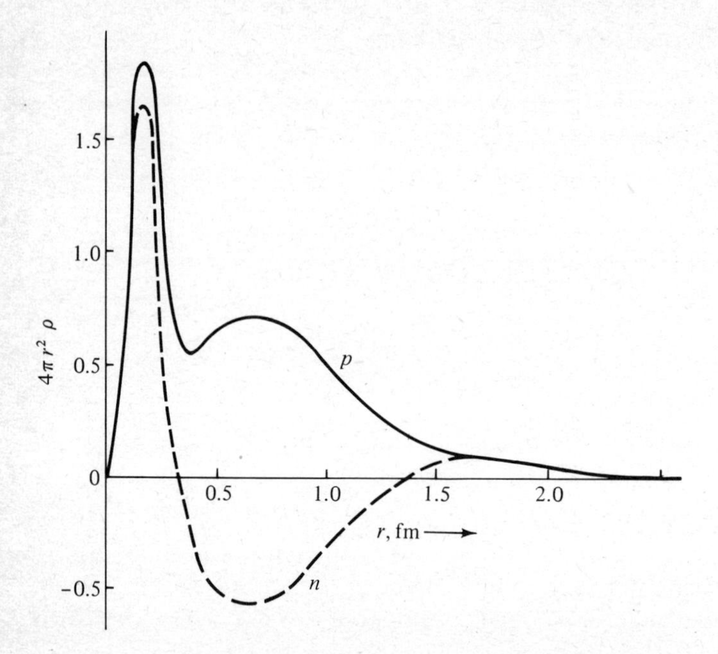

Figure 5-8 Possible charge distributions for the proton and neutron. The ordinate $4\pi r^2\rho$ is proportional to the amount of charge in a shell at radius r. [*From D. Olson, H. Schopper, and R. Wilson, Phys. Rev. Letters* **6**, 286 (1961).]

The intermediate meson V must have the same quantum numbers as the photon, that is, spin 1^-, zero charge, and zero strangeness. There are three such resonances known, the ρ^0, ϕ^0, and ω^0 (Table 3-1). It is interesting to note that long before any of the meson resonances were discovered, Nambu[1] had proposed the existence of a heavy neutral meson to account for the proton and neutron form factors.

In addition to the study of elastic e-N scattering, considerable effort has gone into the study of inelastic scattering

$$e + N \rightarrow e + N'$$

where N' stands for any state with the same quantum numbers as the nucleon but with a greater effective mass. In particu-

[1] Y. Nambu, *Phys. Rev.* **106,** 1366 (1957).

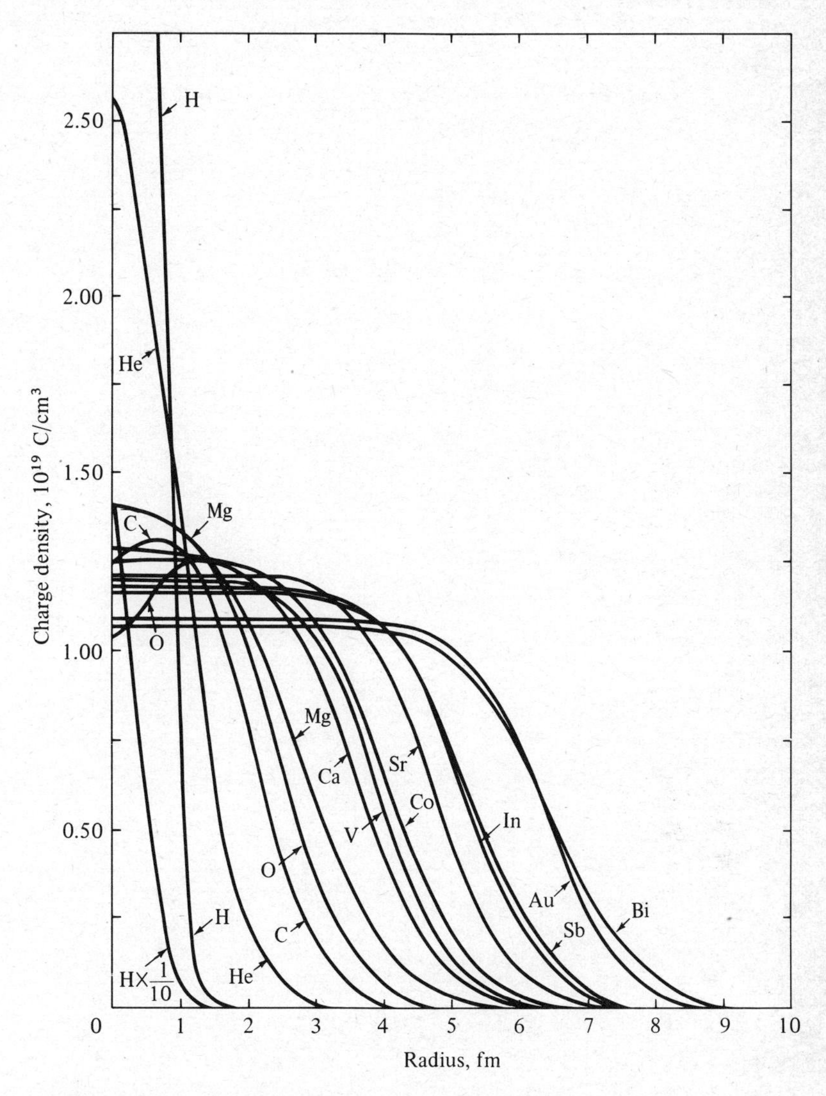

Figure 5-9 Charge distributions for various nuclei as determined from electron-scattering experiments. [*From R. Hofstadter, Ann. Rev. Nucl. Sci.* **7**, 231 (1957).]

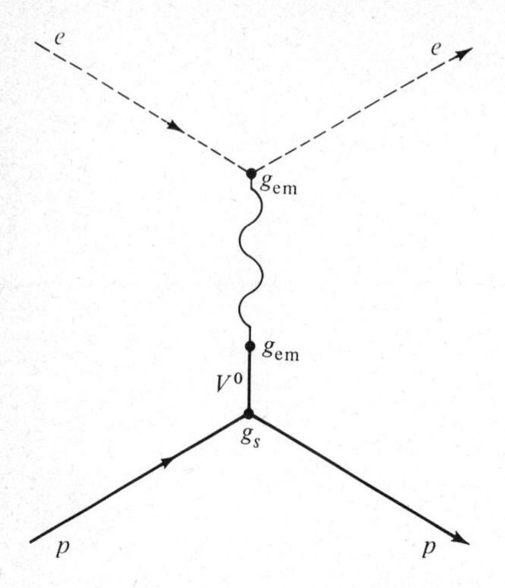

Figure 5-10 Elastic *e-p* scattering with exchange of a neutral, spin 1 meson V^0.

lar, N' can be a nucleon isobar. Figure 5-11 shows some typical data. Peaks in the cross section corresponding to the $\Delta(1236)$, the $N^*(1520)$, and the $N^*(1688)$ stand out clearly. For a given effective mass the cross sections can be fitted to a formula similar to the Rosenbluth formula. The resulting form factors can in principle be used to study the structure of the nucleon isobars. It

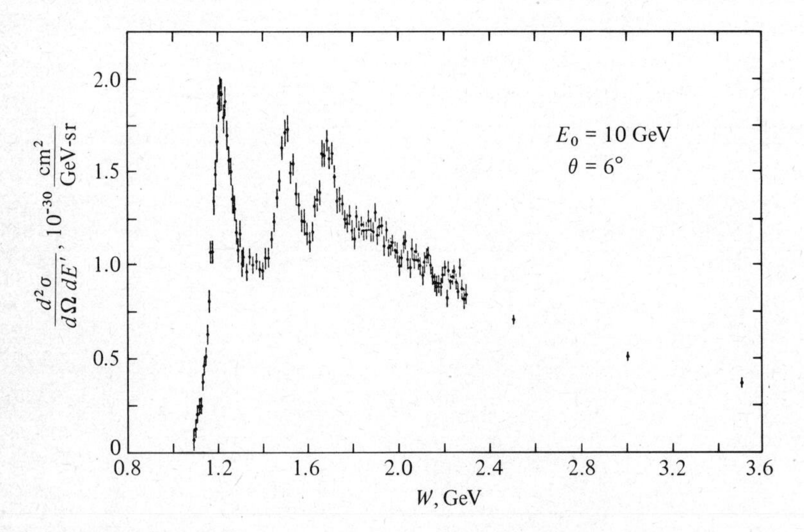

Figure 5-11 Typical inelastic electron-scattering data. W is the effective mass of the N'. (*From Z. Guiragossian, SLAC Rep.* PUB 694.)

is amusing to realize that these experiments are completely analogous to the Franck-Hertz experiment in atomic physics—at 1 billion times the energy!

For very inelastic scattering, when the mass of the N' is well above the known resonances, the form factors seem to become almost independent of q^2. This has been interpreted as possibly indicating the nucleon is made up of several pointlike constituents.[1] The full significance of this result, however, remains to be seen.

5-3 THE MUON

In addition to the electron-scattering experiments, similar experiments have been done with muon beams. The available muon-beam intensities are many orders of magnitude less than those of electron beams, and the statistical errors in the data are therefore larger. The diagram in Fig. 5-1 should also describe muon scattering, *if* the muon has no structure, or in other words, if there is no "box" at the $\mu\gamma$ vertex. Thus the Rosenbluth formula should apply to muon scattering with the same form factors found for electron scattering. Figure 5-12 compares the form factors obtained for muon and electron scattering on protons. The agreement is good, and in fact all experiments to date show that the muon behaves exactly like a "heavy" electron.

This raises some fundamental questions: Why is there a muon? Why is it so much more massive than the electron? Although no one has been able to make a convincing calculation of elementary particle masses, physicists believe that the masses of the particles are somehow generated by their interactions. This view is in the spirit of the classical calculation of the "radius of the electron" by equating the rest mass energy of the electron with the electric potential energy of a sphere of uniform charge density. Classically this energy resides in the electric field. The quantum field theory equivalent is that the rest mass energy resides in the photon field around the electron, as represented in *self-energy diagrams* such as in Fig. 5-13*a*. Similarly the hadron masses are thought to result from analogous diagrams, such as that in Fig. 5-13*b*, which involve

[1] W. Kendall and W. Panofsky, *Sci. Am.* **224,** 60 (June, 1971).

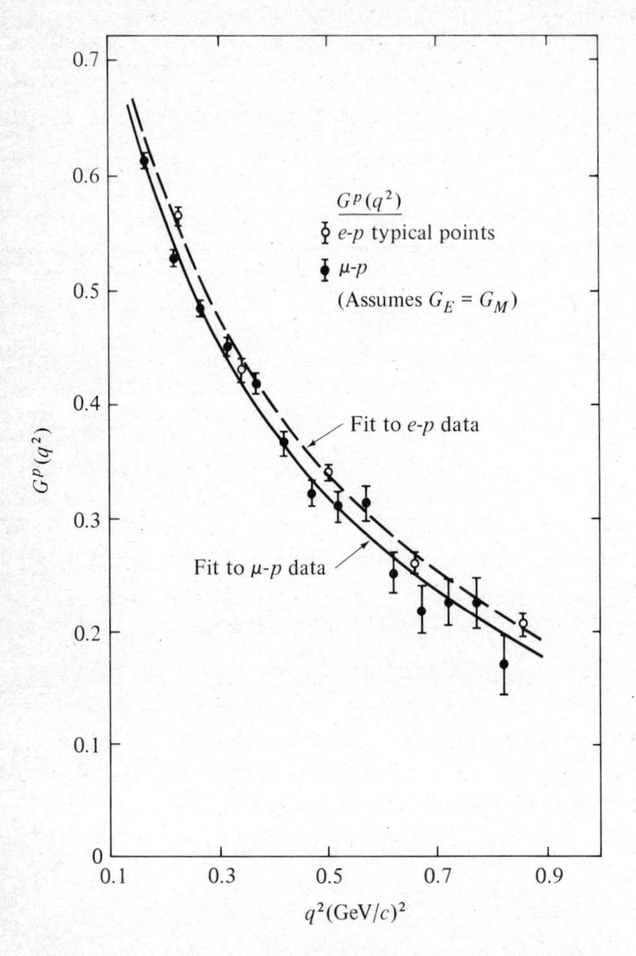

Figure 5-12 Comparison of proton form factors as determined from μ-p and e-p scattering. To facilitate the comparison, it is assumed that $G_E{}^p = G_M{}^p$, in agreement with the electron data. The small systematic discrepancy is probably not significant. [*From Camilleri et al., Phys. Rev. Letters* **23**, 153 (1969).]

strongly interacting particles. The mass splitting between members of the hadron charge multiplets (such as the proton and neutron) is believed to be due to diagrams like those in Fig. 5-13*b*, but involving photons. We therefore expect that if the electromagnetic interactions could be "turned off," the mass splitting between the proton and neutron would disappear, and the electron mass would vanish.

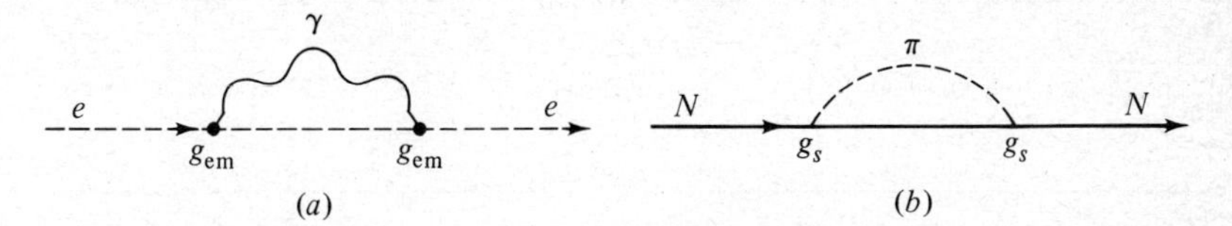

Figure 5-13 (*a*) Electromagnetic self-energy diagram for the electron. (*b*) Self-energy diagram of the sort believed to be mainly responsible for the hadron masses.

It is clear that the muon stands out like a "sore thumb" in this picture. It appears to experience the same electromagnetic interactions as the electron, so that its mass should be the same; yet we find its mass to be ≈ 265 times greater than that of the electron. The puzzle of the muon mass is perhaps the biggest in elementary particle physics.

One has the feeling that the core of the muon must be quite different from that of the electron. If the electron or muon had structure (i.e., were not a point particle), the predictions of quantum electrodynamics which assume a point electron and muon would fail at some level. The two foım factors in the Rosenbluth formula are essentially free parameters and could camouflage a structure in the electron. However, if the muon were different from the electron, we should expect this to show up as a difference between form factors found in e-p and μ-p scattering at large q^2 (corresponding to small distances).[1] So far no differences have been found, and upper limits for the muon radius are $\sim 5 \times 10^{-15}$ cm.

5-4 PHOTOPRODUCTION

It is possible to produce essentially any particle with a photon (γ) beam of sufficiently high energy. Nonstrange mesons or meson-hyperon pairs can be photoproduced singly through reactions such as

$$\begin{aligned} &\gamma + N \to \pi + N \\ &\gamma + N \to V + N \qquad V = \rho,\ \omega,\ \text{or}\ \phi \\ &\gamma + N \to K + \Lambda \end{aligned}$$

[1] Classically for the scattering of pointlike charged particles, the impact parameter b varies approximately as $1/q$.

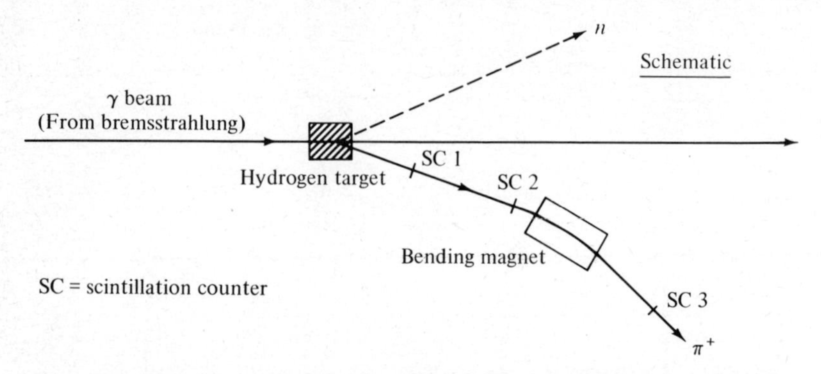

Figure 5-14 Experimental arrangement for studying π^+ photoproduction. The γ beam is usually produced by bremsstrahlung of an electron beam in a high-Z target.

Other particles can be photoproduced in pairs

$$\begin{aligned}\gamma + N &\to e^+ + e^- + N \\ &\to \mu^+ + \mu^- + N \\ &\to p + \bar{p} + N \\ &\to \cdots\end{aligned}$$

The experiments are rather difficult since the cross sections are small and it is usually not possible to obtain monochromatic γ beams. A typical experimental arrangement is shown in Fig. 5-14. The γ beam is usually derived from an electron beam by bremsstrahlung. Recently it has become possible to obtain polarized high-energy γ beams from coherent bremsstrahlung of an electron beam scattered from a diamond crystal[1] or by 180° scattering of an intense laser beam from an oncoming high-energy electron beam.[2]

Single-pion photoproduction, the first of the reactions listed above, has been studied most extensively. The leading diagrams for this process are shown in Fig. 5-15. The observed total cross

[1] H. Überall, *Phys. Rev.* **107,** 223 (1957); G. Diambrini-Palazzi, Monochromatic and Polarized Beams of γ Quanta, in "Proceedings of the 1967 International Symposium on Electron and Photon Interactions at High Energies," CFSTI, Natl. Bur. Stand. (U.S.), Springfield, Va.

[2] R. H. Milburn, *Phys. Rev. Letters* **10,** 75 (1963); F. R. Arutyunian and V. A. Tumanian, *Phys. Letters* **4,** 176 (1963); Diambrini-Palazzi, *loc. cit.*

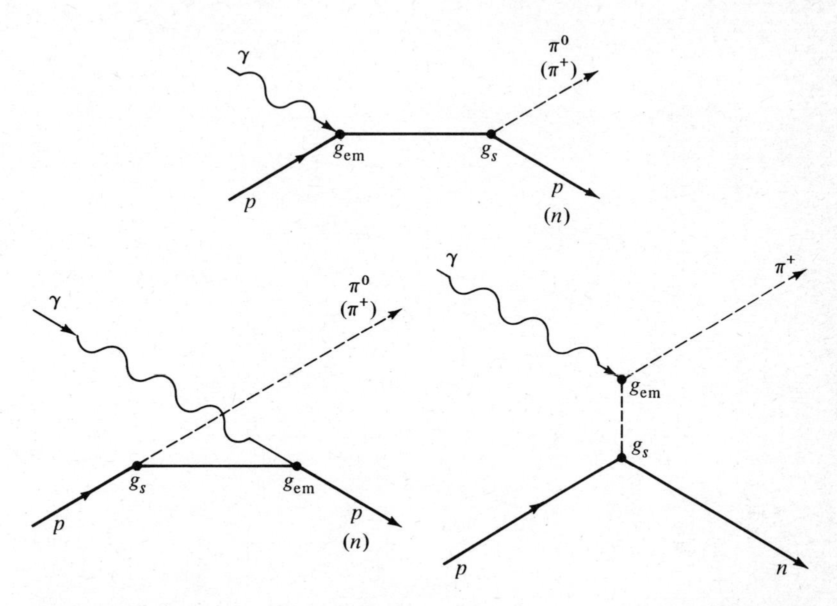

Figure 5-15 Leading diagrams in pion photoproduction. The diagram on the lower right is for charged pions only.

section for π^+ production from protons is shown in Fig. 5-16, plotted versus total energy in the c.m.s. Peaks corresponding to the $\Delta(1236)$, the $N^*(1520)$, and the $N^*(1688)$ show up clearly.

The photoproduction diagrams shown in Fig. 5-15 contain one electromagnetic vertex, and we should therefore expect the photoproduction cross sections to be $\sim\frac{1}{137}$ those for pion scattering. They are actually somewhat smaller. The theoretical analysis of photoproduction is complicated by the fact that isotopic spin is not conserved in electromagnetic interactions. As in N-N scattering, the high angular-momentum states in photoproduction (corresponding to large impact parameters) can be understood in terms of one-pion exchange.

An interesting application of the photoproduction of the vector mesons ρ, ω, and ϕ off of nuclei is in the determination of total cross sections of these particles on nucleons. These can be determined indirectly by studying the dependence on atomic weight of the production cross section near 0° for the reaction

$$\gamma + A \to V^0 + A$$

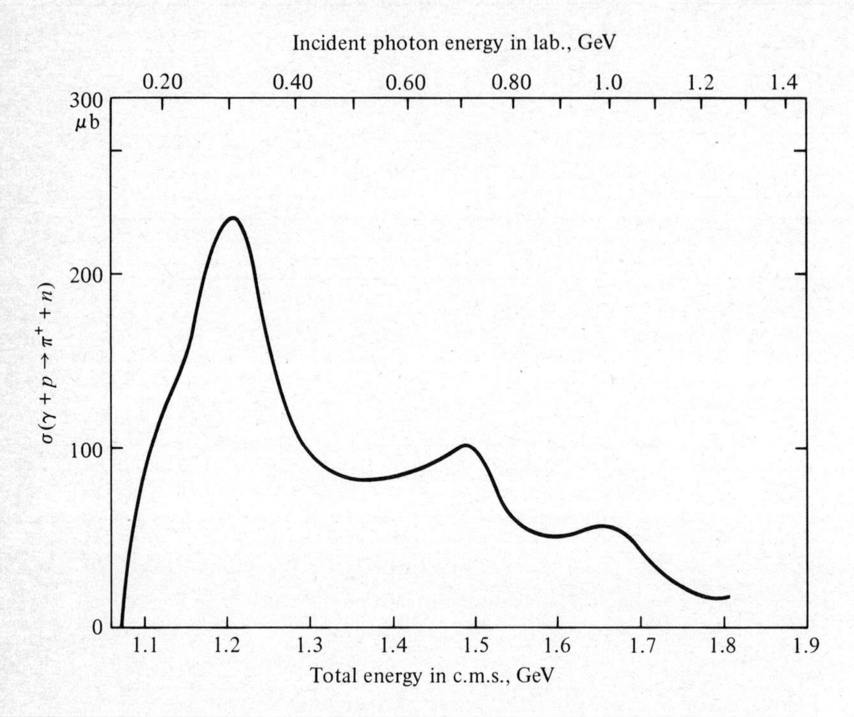

Figure 5-16 Total cross section for $\gamma + p \to \pi^+ + n$ versus energy. *(From J. Beale, S. Ecklund, and R. Walker, Cal. Tech. Rep. CTSL 42, 1966.)*

where A represents a nucleus and V^0 is a neutral vector meson ρ^0, ω^0, or ϕ^0. We shall not consider the analysis in detail; however we can understand the general idea by referring to Fig. 5-17. This shows schematically a high-energy photon incident on a nucleus. The nucleus is relatively transparent for γ's, but the absorption of

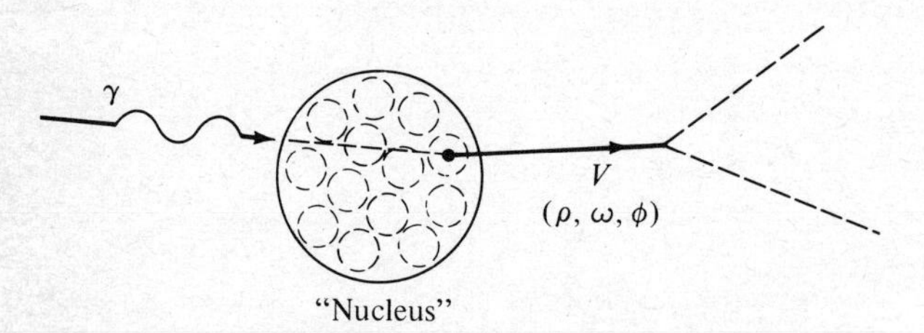

Figure 5-17 Vector meson photoproduction in the coulomb field of a nucleus.

hadrons is usually significant. If the total cross section σ_{VN} for the interaction of the vector meson with nucleons (averaged over neutrons and protons) is small, all the nucleons contribute to the photoproduction near the forward direction. On the other hand, if σ_{VN} is very large, vector mesons produced in the interior of the nucleus will be absorbed and only the layer of nucleons on the downstream edge of the nucleus will contribute to the production cross section near 0°. The dependence of the production cross section on atomic weight is therefore related to σ_{VN}. The available data are still quite crude. They indicate that for energies of several billion electron volts, $\sigma_{\rho N} \approx 29$ mb, $\sigma_{\phi N} \approx 13$ mb, and $\sigma_{\omega N} \approx 27$ mb.

5-5 TESTS OF QUANTUM ELECTRODYNAMICS

Despite its great predictive ability for electromagnetic processes, quantum electrodynamics has its difficulties. For example, a calculation of the electron mass leads to an infinite result. This is patched up in a rather ad hoc way by "renormalizing" this term to the experimental value of the mass. Partially because of such difficulties, physicists have been predicting the imminent failure of quantum electrodynamics for decades. Indeed there have been many instances of experiments which yielded results in serious disagreement with the theory. These have so far always been caused by errors, either in the data or the calculations. In this section we briefly review some of the more significant tests of quantum electrodynamics.

MUON-SCATTERING EXPERIMENTS

As previously discussed in Sec. 5-3, μ-p scattering should be described by the Rosenbluth formula with the form factors found in electron scattering. A discrepancy could be interpreted as due to a nonzero radius for the muon *or* to a breakdown of quantum electrodynamics. The agreement between the e-p and μ-p results is quite good considering the experimental difficulties. It is hard to characterize a nonviolation of the theory since we have no way of predicting how it will fail. Conventionally this is done by introducing a parameter Λ with dimensions of four-momentum transfer. For our purpose it is more convenient to define a parameter $L = \hbar/\Lambda$ with dimensions of length. Crudely speaking, this can be considered a fundamental length within which the theory breaks down. From the muon-scattering experiments it can be concluded that $L \lesssim 10^{-14}$ cm.

ELECTRON-POSITRON-SCATTERING EXPERIMENTS

These experiments can only be done at center-of-mass energies large enough to be of interest with colliding electron-positron beams (Sec. 2-1). This involves storing counterrotating positron and electron beams of very high intensity in the same ring and allowing the beams to collide head on.[1] The simplest diagrams for e^+-e^- scattering are shown in Fig. 5-18. Since these do not include any strongly interacting particles, the cross section can be calculated with no arbitrary constants or form factors. Again the experimental results agree with the theory, indicating a fundamental length

$$L \lesssim 5 \times 10^{-15} \text{ cm}$$

[1] For details of the experimental technique, see W. C. Barber, B. Gittleman, G. K. O'Neill, and B. Richter, *Phys. Rev.* **16,** 1127 (1966).

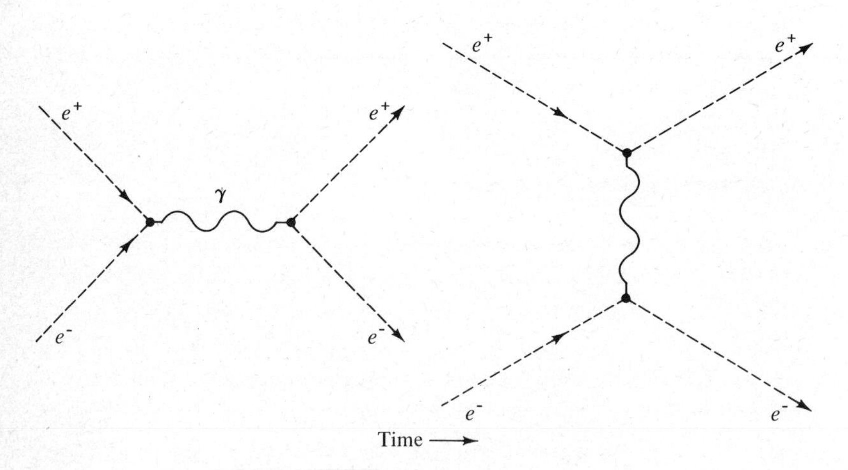

Figure 5-18 Simplest diagrams for e^+-e^- scattering.

THE g FACTOR OF THE MUON

What is often considered the most sensitive test of quantum electrodynamics involves the measurement of the magnetic moment of the muon. If μ is the magnetic moment of the muon in units of $e\hbar/2m_\mu c$, and s its spin angular momentum in units of $\hbar$, the g factor is defined as $g = \mu/s$. Since $\mu \approx e\hbar/2m_\mu c$ and $s = \frac{1}{2}$, then g is very nearly 2. The value of g can be predicted very accurately from quantum electrodynamics with the result that

$$\begin{aligned} a &\equiv \tfrac{1}{2}(g - 2) \\ &= \frac{\alpha}{2\pi} + 0.76578\,\frac{\alpha^2}{\pi^2} + (27 \pm 3)\,\frac{\alpha^3}{\pi^3} + \cdots \end{aligned}$$

where α is the fine-structure constant [$\alpha^{-1} = 137.03608\ (26)$].

The factor $a = \frac{1}{2}(g - 2)$ can be measured directly by observing the precession of the polarization vector of a muon beam moving in a magnetic field. If a muon describes a circular orbit in a magnetic field, it can be shown that the spin vector precesses $(1 + \gamma a)$ times as fast as the momentum vector, where $\gamma = (1 - \beta^2)^{-\frac{1}{2}}$. The direction of the muon polarization as a function of time can be determined by observing the asymmetry of the electrons from the muon decay, $\mu \rightarrow e + \nu + \bar{\nu}$. The higher-energy electrons from μ^- decay tend to come off antiparallel to the muon spin (Sec. 6-4). Thus the counting rate of the decay electrons at a given angle is modulated at a frequency proportional to a.

In the most recent experiment[1] at CERN, polarized muons were formed from π decay (Fig. 5-19). The muons were trapped in a storage ring, and the precession of the muon spin was studied. The observed time distribution of the decay electrons is shown in Fig. 5-20. The result can be written

$$a_{\text{exp}} = \frac{\alpha}{2\pi} + 0.76578\,\frac{\alpha^2}{\pi^2} + (49 \pm 25)\,\frac{\alpha^3}{\pi^3}$$

[1] J. Bailey et al., *Phys. Letters* **B28,** 287 (1968). It is interesting to note that this experiment also provides an experimental confirmation of the relativistic twin paradox to within ~1 percent of accuracy. The muons, traveling with $\gamma \approx 12$ around the ring, undergo constant radial acceleration and are brought back to their starting point. As expected, the muons are found to have a lifetime $\approx$12 times that of a muon at rest in the laboratory.

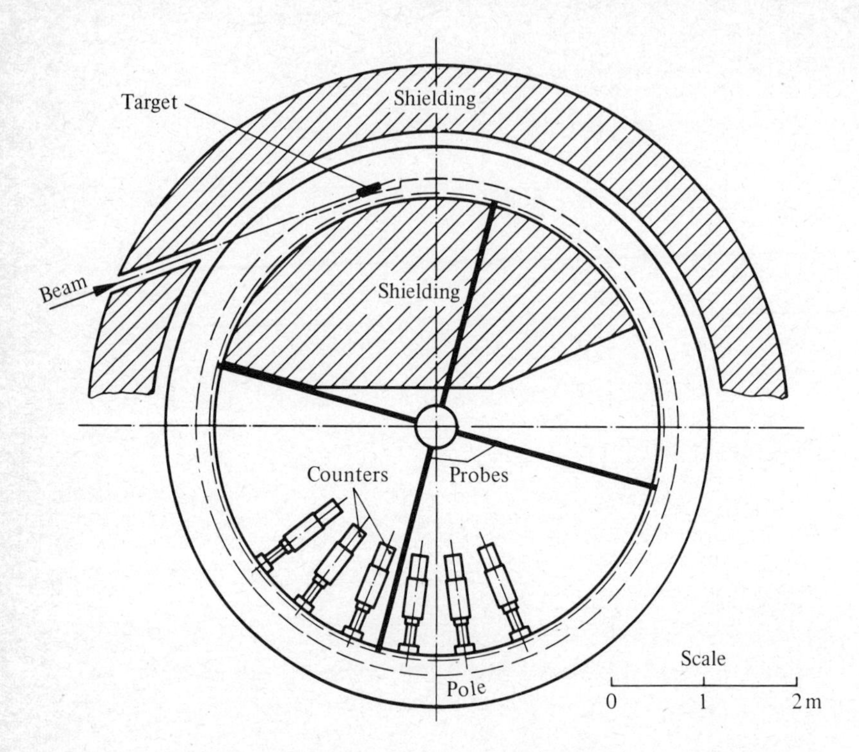

Figure 5-19 Apparatus of Bailey et al. for measuring the g factor of the muon. A short burst of protons from the CERN PS enters from the left and strikes the target. Muons from π-μ decay with a momentum of 1.28 GeV/c are captured in the ring. (There is a magnetic field normal to the plane of the drawing.) The counters detect electrons from μ-e decay. The muons are produced with almost complete longitudinal polarization (see The Two-component Neutrino Hypothesis, Sec. 6-5), and their spin vectors precess in the plane of the drawing as they go around the ring. [*From J. Bailey et al., Phys. Letters* **B28**, 287 (1968).]

The experimental result is thus in satisfactory agreement with the theoretical prediction. Interpreting this in terms of a fundamental length, we find

$$L \lesssim 2 \times 10^{-15} \text{ cm}$$

From arguments based on the radial dependence of the magnetostatic potential of the earth, we know that quantum electrodynamics is valid out to distances of $\gtrsim 5 \times 10^{10}$ cm. Thus quantum electrodynamics appears to remain valid over an incredible range of dis-

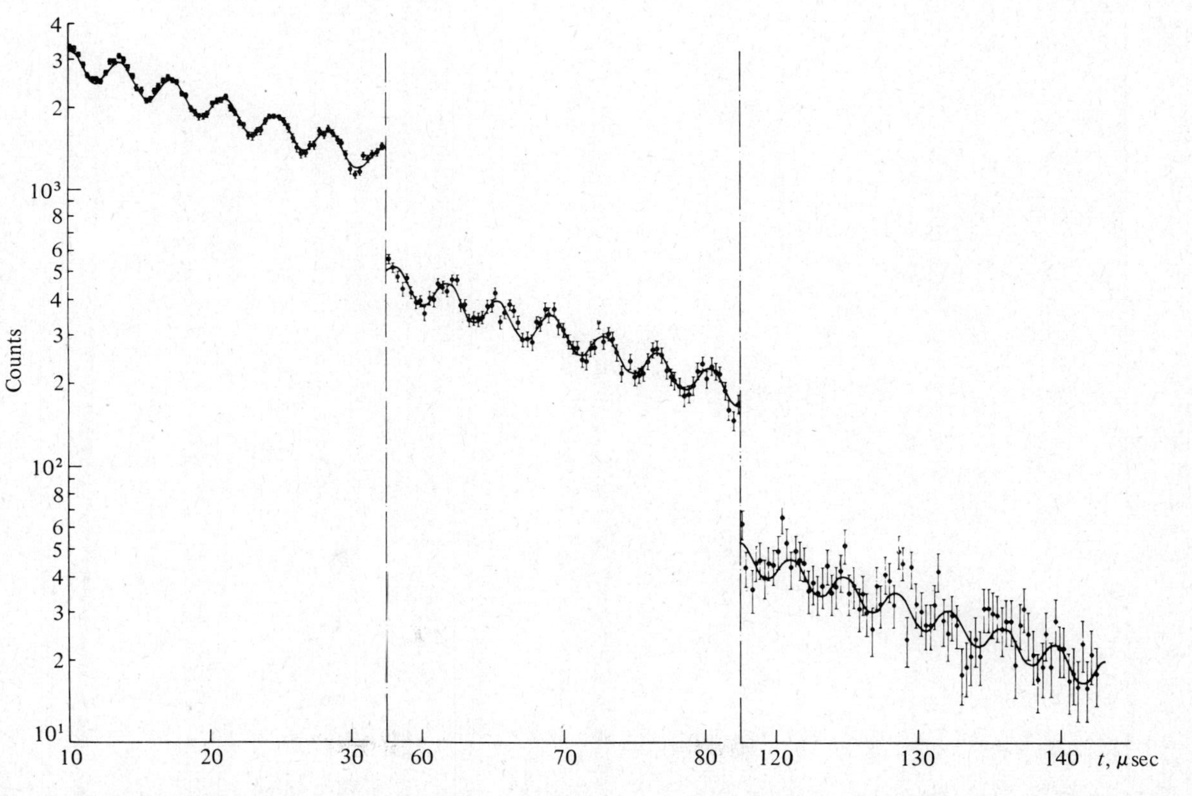

Figure 5-20 Counting rate versus time for electrons from μ decay. The rate is modulated at a frequency proportional to $g - 2$. [*From Bailey et al., Proceedings of the Third International Symposium on Electron and Photon Interactions at High Energies, SLAC* (1967).]

tances, with the range of validity of the theory spanning more than 25 orders of magnitude!

PROBLEMS

5-1 Calculate the differential cross section for 1-GeV e-p scattering from the Rosenbluth formula for $\theta = 12°$ and $\theta = 60°$. Compare with σ_{point}. Repeat for scattering off neutrons.

5-2 A reasonably good approximation to the charge distribution of the proton is

$$\rho(r) = \rho_0 e^{-r^2/a^2}$$

with $a \approx 0.70 \times 10^{-13}$ cm. Use Eq. (5-4) to obtain $F(q^2)$, and compare with the form factors in Fig. 5-7.

5-3 Repeat the above for

$$\rho(r) = \rho_0 e^{-r/a}$$

with $a = 0.8 \times 10^{-13}$ cm.

5-4 When screening of the atomic electrons is taken into account, the potential $U(r)$ for a point nucleus of charge Ze can be approximately represented as

$$U(r) = \frac{Ze^2}{r} e^{-r/r_0}$$

where $r_0 \sim 10^{-8}$ cm. Use the Born approximation (Prob. 3-15) to derive an expression for the differential cross section for electron scattering. Show that in the limit $r_0 \to \infty$, it reduces to σ_{point} if $M \gg E_0$ (so that nuclear recoil can be neglected).

5-5 Derive Eq. (5-3).

5-6 Estimate the mean free path for a high-energy photon in a nucleus if the total cross section is on the order of that for pion photoproduction (Fig. 5-16). Compare this with the diameter of a large nucleus.

CHAPTER 6
weak interactions

The weakness of the weak interactions makes it difficult to study them experimentally. As a result they are often masked by the strong or electromagnetic interactions. Most of the experimental information comes from studies of weak decays. Except for neutrinos, scattering due to weak interactions is negligible compared to that caused by the strong or electromagnetic interactions.

6-1 INTRODUCTORY REMARKS

The best-known weak interaction is β decay, which for a free neutron is (Fig. 6-1)

$$n \to p + e^- + \bar{\nu}$$

Note that the diagram is drawn as though all four particles acted at a *point*. This is significant and is done because, as far as we now know, the weak interaction has zero range. The experimental limit on the range is $\lesssim 2 \times 10^{-14}$ cm. This means, in particular, that only states with zero angular momentum ($\ell = 0$) are involved in weak interactions that do not include hadrons.

Other examples of weak decays are

$$\begin{aligned} \pi^\pm &\to \mu^\pm + \nu\ (\bar{\nu}) \\ K^\pm &\to \mu^\pm + \nu\ (\bar{\nu}) \\ K^0 &\to \pi^+ + \pi^- \\ \Lambda^0 &\to p + \pi^- \\ &\cdots \end{aligned}$$

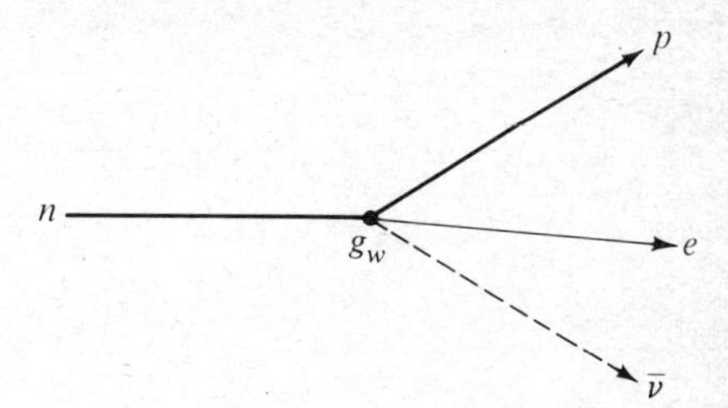

Figure 6-1 β decay of the neutron.

6-2 THE SIX BASIC COUPLINGS

It appears that all the known weak processes can be interpreted in terms of a relatively few basic couplings between particles. These couplings all involve four fermions, for example, $npe\nu$ as in Fig. 6-1. With the help of the virtual strong interactions it is possible to account for all known weak processes in terms of six basic four-fermion couplings. Examples are shown in Fig. 6-2.

The diagrams shown are not unique; there are many other possibilities for the decays in question. The point is that for all observed weak processes we can draw diagrams with a single weak vertex involving four fermions and an arbitrary number of strong vertices. As discussed in Chap. 4, the transition rate will be $\sim g_w{}^2$, no matter how many strong vertices there are.

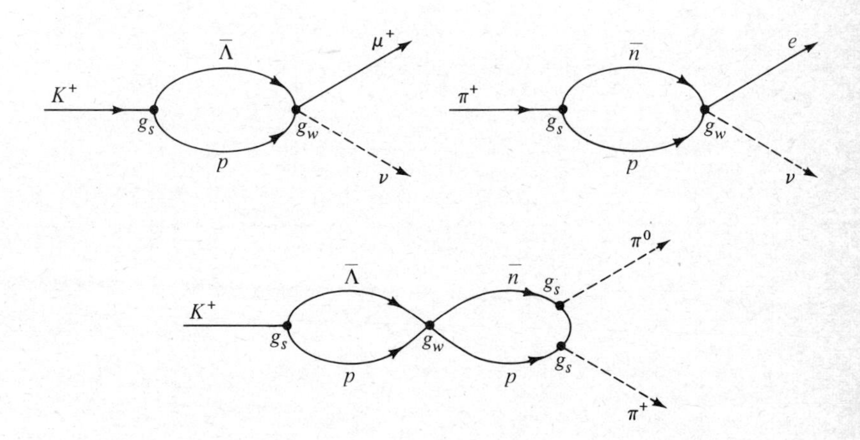

Figure 6-2 Examples of weak decays through four-fermion couplings.

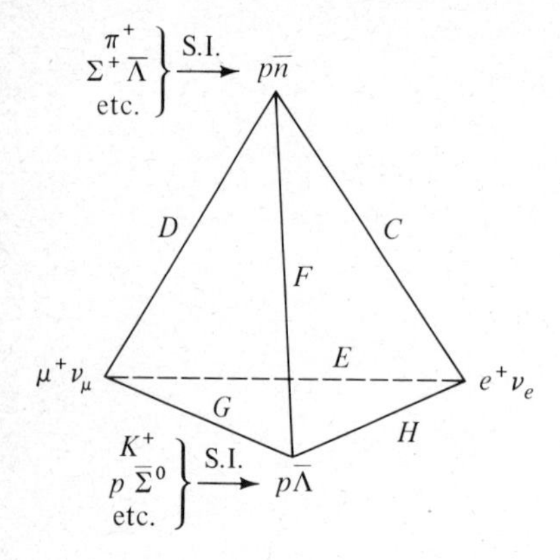

Figure 6-3 The Puppi tetrahedron. The six postulated four-fermion couplings form the legs of the tetrahedron.

It is convenient to array the possible four-fermion couplings on a tetrahedron as shown in Fig. 6-3. (A charge-conjugate tetrahedron is also postulated to account for reactions with negative particles.) Each leg of the tetrahedron corresponds to a possible vertex in a weak interaction. For example, leg E corresponds to the vertex shown in Fig. 6-4. Depending on which are the ingoing lines, this vertex gives reactions such as

$$\mu^+ \to e^+ + \bar{\nu}_\mu + \nu_e$$
$$\bar{\nu}_\mu + e^+ \to \mu^+ + \bar{\nu}_e$$
$$\cdots$$

The tetrahedron of Fig. 6-3 serves as a useful means of summarizing much of what is known about the weak interactions, although its physical significance is not yet clear.

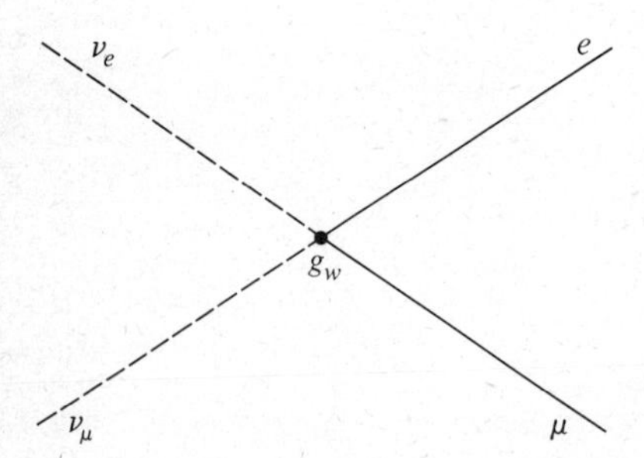

Figure 6-4 The $\mu e \nu_e \nu_\mu$ coupling.

Some notable properties of the weak interactions built into the tetrahedron are:

1. Strange particles are coupled in at only one vertex of the tetrahedron. The three legs F, G, H all involve a strangeness change of *one* unit, while the others involve a strangeness change of zero. Therefore, in general for a *first-order* weak process,

$$\Delta S = 0 \qquad \text{or} \qquad \Delta S = \pm 1 \tag{6-1}$$

   Thus decays such as $\Xi^0 \rightarrow n + \pi^0$ are forbidden to occur as a first-order weak process. (They may occur through a second-order weak interaction, but their rates would then be extremely small and such decays have never been observed.)
2. Reactions involving two or more neutrinos, such as $\pi^+ \rightarrow \mu^+ + 3\nu$, $K \rightarrow \pi + \nu + \bar{\nu}$, and $\nu + p \rightarrow \nu + p$, are not included, except for $\mu \rightarrow e + \nu_e + \nu_\mu$. Such processes are presumably forbidden, at least in first-order weak processes.
3. Experimentally it is found that the strength of the coupling along the six legs of the tetrahedron is approximately equal. This is the basis of the hypothesis of the *universal Fermi interaction.* It is difficult to give a precise definition of this hypothesis at this level, but roughly speaking, it corresponds to saying that in the absence of strong interactions[1]

$$g_w{}^C = g_w{}^D = g_w{}^E = g_w{}^F = g_w{}^G = g_w{}^H \tag{6-2}$$

   This is analogous to finding that all electromagnetic interactions have the same coupling constant $\sqrt{\frac{1}{137}}$, regardless of whether we are talking about μ's, e's, p's, Σ's, or other particles. (It did

[1] Two coupling constants are actually needed, one corresponding to a vector coupling and the other to an axial-vector coupling. These are apparently equal in magnitude but opposite in sign. As in the electromagnetic interactions, the strong interactions complicate weak-interaction theory considerably for processes involving hadrons. In particular, the rates for strangeness-violating weak decays (such as $K^+ \rightarrow \mu^+ + \nu_\mu$) are about 30 times slower than expected from the rates of corresponding strangeness-conserving reactions (such as $\pi^+ \rightarrow \mu^+ + \nu_\mu$). A procedure for modifying the coupling constants has been proposed by N. Cabibbo. Weak-interaction theory now seems capable of predicting rates and other data for a large variety of weak processes involving hadrons. (See, for example, C. S. Wu and S. A. Moszkowski, "Beta Decay," Interscience-Wiley, New York, 1966.)

not have to work out that way, and we may yet find that the equality is not exact.)

It is not known for certain whether the vertices of the tetrahedron are coupled to themselves, for example, in reactions such as $\nu_e + e^{\pm} \rightarrow \nu_e + e^{\pm}$ and $e^+ + e^- \rightarrow \nu_e + \bar{\nu}_e$. It is believed however that the former reaction is important in star cooling. There is also good indirect evidence for a $p\bar{n}p\bar{n}$ interaction from studies of parity-violating effects in nuclear γ decay.

We have also incorporated in the tetrahedron another experimental result, the existence of two distinct neutrinos ν_e and ν_μ. The recognition of this was forced by the fact that the reaction $\nu + n \rightarrow e^- + p$ occurs with neutrino beams which are produced in conjunction with electrons through various β decays in reactors. On the other hand, neutrino beams which are produced in conjunction with muons [mostly in π decays, $\pi \rightarrow \mu + \nu$ (see Sec. 6-5)] do not produce the above reaction, but instead produce *muons* through the reaction

$$\nu + n \rightarrow \mu^- + p \qquad \sigma \sim 10^{-38}\ \text{cm}^2$$

Some peculiarity of electrons compared with muons had already been suspected, since the reaction $\mu^{\pm} \rightarrow e^{\pm} + \gamma$ is "possible" but has never been observed. The situation can be described in a somewhat ad hoc way by invoking a new quantum number and corresponding conservation law. This can be done by assigning a *muon number* L_μ and an *electron number* L_e to all particles according to the scheme in Table 6-1. It is then supposed that L_e and L_μ are separately

Table 6-1 Muon-number and electron-number assignments to the particles

Particle	L_μ	L_e
μ^+, $\bar{\nu}_\mu$	−1	0
μ^-, ν_μ	+1	0
e^+, $\bar{\nu}_e$	0	−1
e^-, ν_e	0	+1
All other particles	0	0

conserved so that ΣL_μ and ΣL_e remain constant for any closed system. Although it is important to keep in mind the distinction between the two kinds of neutrinos, in what follows we shall often omit the subscripts μ and e for simplicity.

6-3 STATUS OF THE THEORY: THE INTERMEDIATE BOSON

It might seem that the weak interactions would be relatively easy to treat theoretically because of their weakness. Nevertheless our understanding of the weak interactions is still quite limited. This is partly because of the effect of virtual strong interactions; note, for example, that all legs of the tetrahedron except E involve hadrons. The other difficulty, basically an experimental one, is that the relevant experiments for the most part range from difficult to impossible. Thus there is a lack of solid experimental data with which to confront the theory, particularly for large energies in the c.m.s. since these can only be studied in neutrino-scattering experiments.

We shall not be able to discuss the theory except in very general terms. The theory does seem capable of describing the purely weak interaction quantitatively. However it is clear that it has significant problems. For example, it breaks down in the high-energy limit, predicting infinite cross sections for neutrino scattering in the limit of infinite energy.

A way out of some of the theoretical difficulties and an attractive possibility in its own right is the hypothesis that the weak interaction is mediated by a particle, the W, whose role is analogous to that of the photon in the electromagnetic interactions. The diagram for neutron β decay would then be as pictured in Fig. 6-5 instead of as shown in Fig. 6-1. At each vertex, $\sqrt{g_w}$ appears, so that the overall amplitude contains g_w.

The characteristics of the postulated W particle can be easily predicted from the known properties of the weak interactions:

1. Both a W^+ and a W^- are needed to account for the couplings shown in Fig. 6-3 and their charge conjugates. A W^0 however need *not* exist, and in fact its existence seems to be excluded

by the very low limits placed on decays such as $K^0 \rightarrow \mu^+ + \mu^-$ (Table 1-1), which would require a neutral intermediate particle.

2. From the upper limit on the range of the weak interactions we can place a *lower* limit on the mass of the W. The argument parallels that used in Sec. 3-3 to relate the range of the strong interaction to the pion mass, and leads to a lower limit of about 1 GeV for the mass of the W.

3. The W is coupled to all the pairs of particles at the vertices of the tetrahedron (Fig. 6-3), and can therefore decay into any of these pairs with a lifetime $\sim g_w^{-\frac{1}{2}}$ times the lifetime for a typical strong decay. The lifetime of the W is therefore expected to be of order[1]

$$\begin{aligned}\tau_W &\sim (10^{-14})^{-\frac{1}{2}}(10^{-24}) \text{ sec} \\ &\sim 10^{-17} \text{ sec}\end{aligned}$$

which is too short to allow it to be observed directly. The W might show up as a rather long-lived (very narrow) resonance with a variety of decay modes, including charged lepton-neutrino pairs.

4. The W must have spin 1. This can be most easily seen by considering β decay, which we assume to occur through the diagram in Fig. 6-5. It is known (see Sec. 6-5) that in β decay the spin of the $\bar{\nu}$ is always parallel to its momentum vector (Fig. 6-6), and that of the e is antiparallel. The spin of the parent W must therefore be 1 to conserve angular momentum. The W would

[1] This is probably on the high side. Since the W is presumably very massive, the available energy for the decay is quite large and the lifetime correspondingly short.

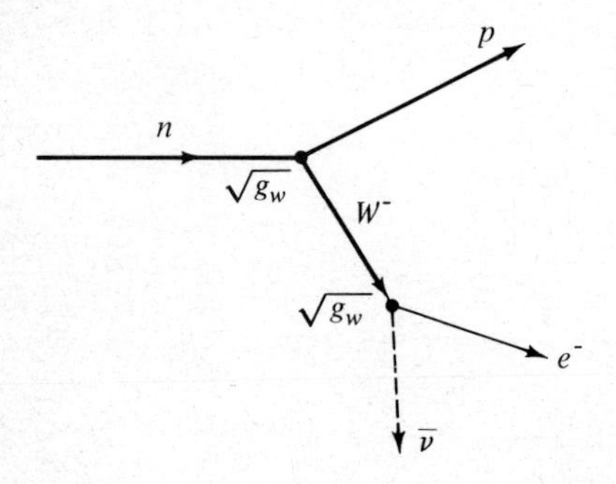

Figure 6-5 Neutron β decay with an intermediate boson.

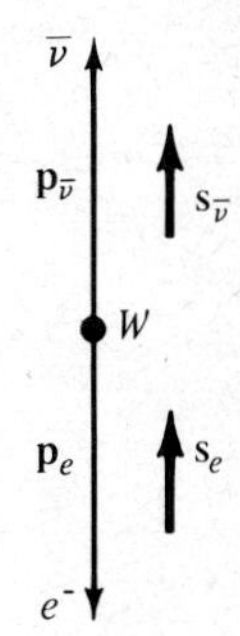

Figure 6-6 The decay of the W^- in its own rest frame. The short arrows represent the spins, and the long arrows the momenta of the e and $\bar{\nu}$. By angular-momentum conservation the W must have a spin of 1.

thus be a boson, and it is often referred to as the *intermediate vector boson.*[1]

If the W exists, predictions based on a point-interaction theory would fail at distances $\lesssim \hbar/m_W c$ or correspondingly at momentum transfers $\sim m_W c$. Many searches for the W have been made, but none have been successful. In principle the W can be produced in any of the interactions it is presumed to mediate, but production cross sections are likely to be rather small since it is produced "semiweakly" with a coupling constant $\sim \sqrt{g_w}$.

The most readily interpreted W search has been made with neutrino beams. The neutrino can virtually dissociate into a μ-W pair, and the W can be "materialized" by allowing it to transfer momentum to some nearby object such as a nucleus. The type of process possible is illustrated in Fig. 6-7. The W can decay in a variety of ways. The reaction sought was

$$\nu_\mu + \text{Nu} \rightarrow \text{Nu} + \mu^- + W^+ \quad \llcorner\!\!\rightarrow \begin{cases} \mu^+ + \nu_\mu \\ e^+ + \nu_e \end{cases} \qquad (6\text{-}3)$$

The experiment therefore consisted in making an intense beam of high-energy neutrinos and looking at the neutrino-induced reactions in a massive spark chamber or bubble chamber. The "signa-

[1] A particle with spin 1 is commonly called a *vector particle* because its wave function has three components (corresponding to the three possible spin orientations), like a vector in three-dimensional space.

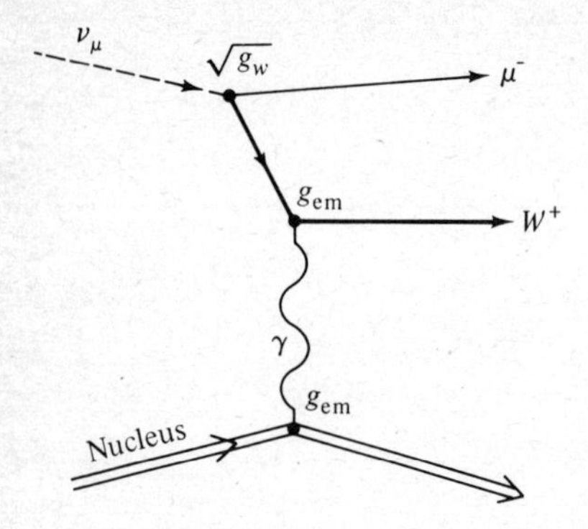

Figure 6-7 Possible diagram for the production of a W^+ by a neutrino. The neutrino dissociates into a μ^- and W^+. Overall momentum-energy conservation is satisfied by transferring momentum to the nucleus by photon exchange.

ture" of a W event is the appearance of two charged leptons at a vertex in the chamber.[1] Since cross sections for producing W's in reaction (6-3) can be reliably calculated as a function of the mass of the W, the failure to observe W's can be interpreted in terms of a lower limit on the W mass. The present limit is $m_W \gtrsim 2$ GeV, based on neutrino experiments at CERN and Brookhaven National Laboratory.

6-4 VIOLATION OF C AND P INVARIANCE IN WEAK INTERACTIONS

As noted previously in Chap. 1, the weak interactions exhibit large violations of symmetry with respect to charge conjugation C and inversion of coordinates, P. An experiment to study β decay in 1928 indicated the possibility of parity violation in the weak interaction. However the general belief that nature was mirror symmetrical led physicists to disbelieve this result. The question was not revived until about 1956. At that time physicists were discovering what seemed to be several particles with masses intermediate between the pion and nucleon. We now know that what were being observed were the complicated decay schemes of the K mesons. In particular there seemed to be two particles with essentially the

[1] This is a convenient signature because charged lepton pairs are rarely produced in other processes. In addition, high-energy muons and electrons leave rather distinctive tracks; the muons show up as long, noninteracting tracks, and the electrons produce showers.

same mass and lifetime but different intrinsic parity, the $\theta^{\pm}$ and $\tau^{\pm}$, with decay schemes

$$\theta^{\pm} \rightarrow \pi^{\pm} + \pi^{0}$$
$$\tau^{\pm} \rightarrow \pi^{\pm} + \pi^{+} + \pi^{-}$$

It was possible to determine the spin-parity of the τ from the distribution in energy and angle of the decay products.[1] It was found that the τ had spin 0^{-}, which we now know to be correct for the kaon.

On the other hand, a particle like the θ which decays into two pions "cannot" have zero spin and odd parity (*assuming* parity conservation). For two pions the overall parity is $(-1)^2(-1)^{\ell} = +1$ if the θ has zero spin, since total angular momentum is conserved in the decay. Therefore the θ "cannot" be 0^{-} and must be distinct from the τ. The fact that it had the same mass and lifetime was presumably accidental.

A possible solution to the τ-θ dilemma was suggested by T. D. Lee and C. N. Yang. They pointed out that parity conservation in the weak interactions had not been tested experimentally. The τ and θ could be the same particle if the decay violated P. As is well known, a series of experiments shortly thereafter showed that there was indeed a large violation of C and P in weak decays.

To illustrate how the C and P violations show up experimentally, let us reconsider the example of the decay of polarized muons,

$$\mu^{+} \rightarrow e^{+} + \nu + \bar{\nu}$$

previously discussed in Chap. 1 (see Fig. 6-8). Symmetry under P means that all observables, in particular the angular distribution of the decay positrons, must be symmetrical upon reflection through the origin, $\mathbf{r} \rightarrow -\mathbf{r}$. (Since the problem considered has complete symmetry relative to rotation about the Z axis, this is equivalent for our purpose to a reflection in the mirror.) Remember that $\mathbf{s}_{\mu}$ is unaffected by P. Symmetry under P therefore requires that the angular distribution of positrons be symmetrical about the XY plane. It cannot, for example, contain a $\cos\theta$ term. In general, P invariance requires that no observable contain terms odd with respect to P. A term in $\cos\theta \propto \mathbf{s}_{\mu} \cdot \mathbf{p}_{e}$ is clearly odd, since $\mathbf{p}_{e}$ is odd and $\mathbf{s}_{\mu}$

[1] See, for example, W. Frazer, "Elementary Particles," chap. 5, Prentice-Hall, Englewood Cliffs, N.J., 1966.

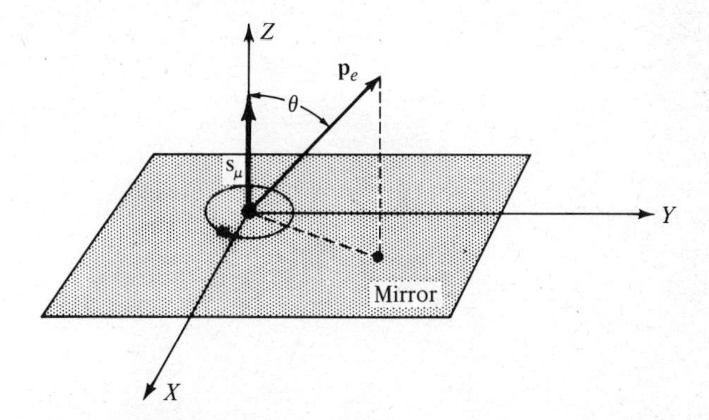

Figure 6-8 The decay of a polarized muon, $\mu \rightarrow e + \nu + \bar{\nu}$. Symmetry with respect to the parity operation requires that the distribution of decay electrons be symmetrical about the XY plane.

is even under P. In the decay of polarized muons it is found experimentally that the angular distribution of the decay electrons is strongly asymmetrical about the XY plane, which indicates a large P violation. This asymmetry in the positron distribution, by the way, provides a very convenient way of measuring muon polarization.

The first decisive experiment to show a P violation was that of C. S. Wu et al.,[1] who showed that the electron distribution from the β decay of polarized Co^{60} was asymmetrical.[2] This corresponds

[1] C. S. Wu, E. Ambler, R. Hayward, D. Hoppes, and R. Hudson, *Phys. Rev.* **105,** 1413 (1957).

[2] The discovery of parity violation moved W. Pauli to write in a famous letter, "God is left-handed!" In fact this question is still unresolved.

T. D. Lee. Born 1926 in Shanghai, China. In 1946 he was awarded a Chinese government scholarship which took him to the University of Chicago. There he did a thesis on white dwarf stars, and for a short time after receiving his Ph.D. he worked at Yerkes Astronomical Observatory. Between 1951 and 1953 he was a member of the staff of the Institute for Advanced Research at Princeton, N.J. There he and C. N. Yang developed the idea that parity was not conserved in weak interactions. In 1957 Lee and Yang jointly received the Nobel Prize for their work on parity nonconservation. In addition to his numerous important contributions to elementary particle physics, Lee has made major contributions to statistical mechanics, nuclear physics, field theory, and astrophysics. (*Photograph from the Niels Bohr Library, American Institute of Physics.*)

C. N. Yang. Born 1922 in Hofei, Anwhei, China. Yang first met Lee at the National Southwest University in K'unming, Yunnan, where both of them studied after the Japanese invasion of China. Yang went to the University of Chicago in 1946 on a Tsinghua University fellowship. He received his Ph.D. in 1948 under the guidance of Professor E. Teller. Yang's chief contributions have been in the fields of statistical mechanics, symmetry principles, and elementary particle theory. Both Lee and Yang are naturalized United States citizens. (*Photograph from the Marshak Collection, Niels Bohr Library, American Institute of Physics.*)

to a term $\mathbf{s}_{\text{Co}} \cdot \mathbf{p}_e$ in the decay distribution. As discussed above, such terms are odd with respect to space inversion and violate parity.

There are many other examples of parity violations in weak interactions. Another interesting one is in the decay of polarized hyperons, for example, Λ^0. It is known that Λ^0's produced in the reaction

$$\pi^- + p \to \Lambda^0 + K^0$$

have a large polarization perpendicular to the production plane (defined by $\mathbf{p}_\pi \times \mathbf{p}_\Lambda$) for pion energies of approximately 1 GeV and certain angular ranges.[1] When the Λ's decay, a large asymmetry of the decay π's, corresponding to a term $\mathbf{s}_\Lambda \cdot \mathbf{p}_\pi$, is observed. This term is odd under P, and thus its existence indicates a P violation.

Violations of C invariance show up when we compare the decay of a particle with that of its antiparticle, for example, μ^+ with μ^-. The prediction of C invariance is that the angular distribution of positrons from the decay of polarized μ^+ should be the same as that of electrons from μ^- decay. In fact, these turn out to be mirror images of each other as illustrated in Fig. 6-9, which schematically represents the $e^\pm$ distributions. The positrons from μ^+ decay come off preferentially along the spin direction, while the electrons from μ^- decay tend to come off opposite to the spin direction. Furthermore the $e^\pm$ from $\mu^\pm$ decays are longitudinally polarized, with the

[1] The Λ polarization is of course the result of strong interactions. It corresponds to a term $\mathbf{s}_\Lambda \cdot (\mathbf{p}_\pi \times \mathbf{p}_\Lambda)$, which is even under P.

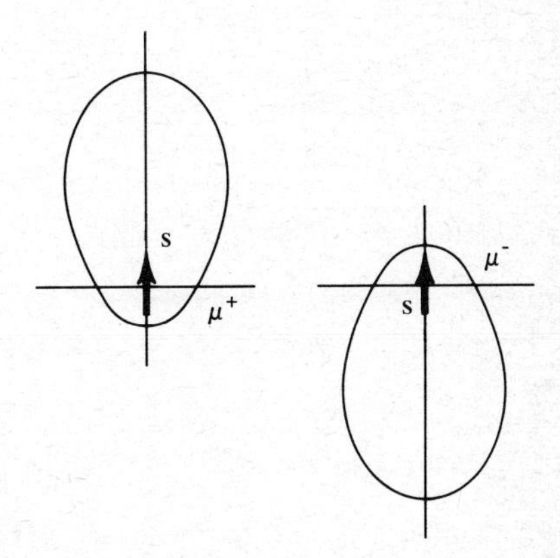

Figure 6-9 Polar plots of the observed decay electron distributions for μ^+ and μ^- decay (schematic). Positrons from μ^+ decay tend to come off along the muon spin direction. Electrons from μ^- decay tend to come off opposite to the muon spin direction.

polarization a function of θ. The existence of such a polarization is another example of a parity violation since it corresponds to a term $\mathbf{s}_e \cdot \mathbf{p}_e$. Again the C violation shows up as a difference in the angular distribution of the polarization in the two cases, the two distributions being mirror symmetrical.

Muon decay nicely illustrates a symmetry which nature does seem to possess (though, as we shall see, it is apparently violated to some extent). This is symmetry with respect to the *combined* operation CP. Thus if we change μ^+ to μ^- and simultaneously reflect coordinates, we get the observed electron distribution in μ^- decay. In general CP invariance requires that if we change each particle in a system to its antiparticle and invert coordinates, we get an equally probable state of the antiparticle system.

It is interesting to note that we can use the parity violation in the weak interactions to communicate to a scientist on some distant planet our definition of a right-handed coordinate system. A possible prescription would be the following:[1]

> Take a system of three mutually perpendicular axes XYZ. At the origin, situate a μ^+ spinning from X to Y (through the smallest angle). If the decay electrons go preferentially along the Z axis, it is a right-handed system.

The hitch is that we have to be sure he is not living in a world of antimatter. We might define μ^+ as that muon with the same charge as the atomic nucleus. However in an antiworld this would be the particle we call μ^-, and our prescription for defining a right-handed system would fail.

6-5 NEUTRINOS AND THEIR PROPERTIES

GENERAL DISCUSSION

The neutrino is certainly the most peculiar particle in the particle zoo. It is (apparently) massless and therefore, like the photon,

[1] It is not hard to think of ways of communicating this information to the other observer directly, e.g., by the use of polarized radio waves. However this violates the rules of the "game," which are to communicate this information without the use of directional cues.

must travel at the velocity of light to exist. As discussed earlier, there are at least two subspecies ν_e and ν_μ with their antiparticles. The existence of the neutrino was first suggested by W. Pauli around 1930 to explain the apparent violation of energy conservation in β decay. Early experiments to detect neutrinos were unsuccessful. The reason they failed is clear as soon as we realize that a neutrino with an energy $\sim$1 MeV, typical of those from β decay, can pass through the earth with negligible chance of interacting! (See Prob. 6-1.) It was not until 1953 that the existence of the neutrino was shown directly by C. Cowan and F. Reines.[1] They observed the reaction $\bar{\nu}_e + p \rightarrow n + e^+$, using a nuclear reactor as an intense source of $\bar{\nu}$'s.

Aside from the study of weak decays, neutrino-scattering experiments are essentially our only other source of information on the weak interactions. This is because the neutrino is apparently the only particle which interacts exclusively through weak (and gravitational) interactions. In the scattering of particles other than neutrinos, weak-interaction effects are invariably masked by the much larger cross sections for strong or electromagnetic interactions.

Neutrino scattering is also the only practical means of studying the weak interaction at high energies, beyond those accessible in weak decays. Such studies at high energies are especially important in view of the divergence of the theory at high energy mentioned previously. In the last few years high-energy neutrino beams of sufficient intensity have become available, and a significant amount of data on neutrino interactions in the energy range up to about 10 GeV has been accumulated. We can estimate the maximum energy for which the present theory (which assumes a *point* interaction of four fermions) is valid by the following argument: In a reaction involving leptons exclusively, such as $\nu_u + e^- \rightarrow \nu_e + \mu^-$, only the $\ell = 0$ state need be considered if a point interaction is assumed. The maximum cross section for $\ell = 0$ is $\frac{1}{2}\pi\lambda\!\!\!^{-2}$, based on Eq. (3-9*a*) and its generalization given in Sec. 3-5.[2] Weak-inter-

[1] C. Cowan and F. Reines, *Phys. Rev.* **90,** 492 (1953); F. Reines, *Ann. Rev. Nucl. Sci.* **10,** 1 (1960).

[2] There is an additional factor of $\frac{1}{2}$ because this is an inelastic channel and diffraction scattering does not contribute.

action theory predicts a cross section for this reaction given by

$$\sigma = \frac{4}{\pi} g_w{}^2 p_\nu{}^2 \qquad (6\text{-}4)$$

where g_w is the weak-interaction coupling constant, and p_ν the neutrino momentum in the c.m.s. The cross section predicted by Eq. (6-4) exceeds the limit $\frac{1}{2}\pi\lambda\!\!\!^{-2}$ for $p_\nu \gtrsim 300$ GeV/c. We are therefore assured of a breakdown in the theory before $p_\nu = 300$ GeV/c. It would be of great interest to observe this breakdown since it might give a clue to how the present theory must be modified. Unfortunately $p_\nu = 300$ GeV/c in the c.m.s. corresponds to $\approx 3.6 \times 10^8$ GeV/c in the laboratory! This makes prospects for seeing a breakdown with present accelerators rather dim. Kinematics for the reaction

$$\bar{\nu}_\mu + p \rightarrow n + \mu^+$$

are somewhat more favorable, but the theoretical cross sections in this case contain unknown form factors to account for the structure of the proton, just as in electron scattering. With certain assumptions, two of the three form factors can be related to those for electron scattering, but the third is still unknown. In any case, progress in this very difficult field is likely to be slow.

NEUTRINO EXPERIMENTS AT ACCELERATORS

High-energy proton accelerators are relatively copious sources of secondary beams of π's, K's, and other relatively long-lived hadrons. Neutrinos are not produced directly in significant numbers, but are produced in weak decays of secondaries, in particular,

$$\pi^\pm \rightarrow \mu^\pm + \nu_\mu\ (\bar{\nu}_\mu)$$
$$K^\pm \rightarrow \mu^\pm + \nu_\mu\ (\bar{\nu}_\mu)$$

Electron neutrinos ν_e are produced much less frequently, mainly in the decay

$$K \rightarrow \pi + e + \nu_e$$

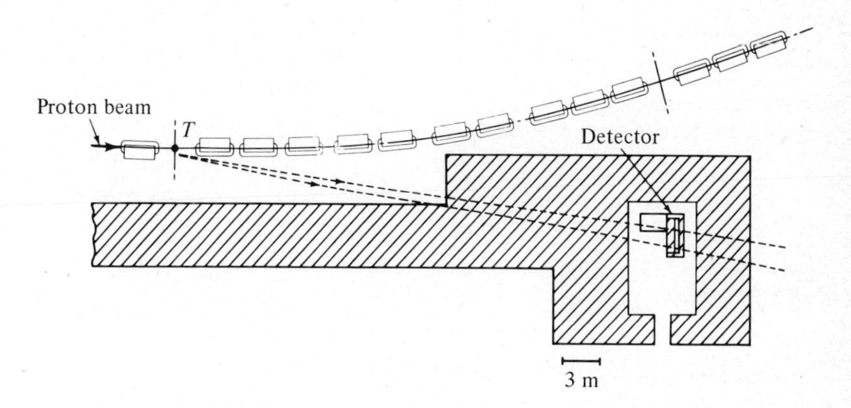

Figure 6-10 Experimental arrangement used by Danby et al. to study neutrino interactions. The shielding was designed to stop essentially all hadrons and charged leptons so that only neutrinos reached the detector. [*From Danby et al., Phys. Rev. Letters* **9**, 36 (1962).]

The main difficulty in neutrino experiments is of course the very small cross sections, which are typically $\sim 10^{-14}$ times those for strong interactions. Neutrino experiments at accelerators therefore only became practical in the 1960s with the advent of high-energy, high-intensity machines (in particular, the Brookhaven AGS and the CERN PS) and the development of very large detectors (first, spark chambers and, more recently, very large bubble chambers).

The first successful neutrino experiment with an accelerator was that of Danby et al.[1] with the Brookhaven AGS. The experimental arrangement they used is shown in Fig. 6-10. The proton beam of the AGS (only a small section of which is shown) was accelerated to about 15 GeV and allowed to strike a beryllium target T inside the vacuum tank. The pions and kaons produced tend to go forward along the direction of the proton beam. A space of approximately 21 m was provided for the pions to decay. Then the remaining pions, muons, and all other particles except neutrinos were stopped by a massive iron shield, about 13 m thick.[2] This was designed to ensure that only neutrinos could reach the detector. The detector consisted of a spark-chamber array with 1-in.-thick

[1] G. Danby, J. M. Gaillard, K. Goulianos, L. M. Lederman, N. Mistry, M. Schwartz, and J. Steinberger, *Phys. Rev. Letters* **9**, 36 (1962).

[2] Much of the steel came from the defunct battleship *Missouri*.

aluminum plates and a total mass of about 10 tons. (A similar array of spark chambers used by the same group in a later experiment is shown in Fig. 6-11.) The neutrino flux at the detector was approximately 1.5×10^7 m^{-2} for each pulse of protons from the AGS, which operated at a rate of about one pulse per second.

The signature of a neutrino event is one or more charged particles produced in the chamber array. One such event is shown in Fig. 6-12. The first experiment yielded a total of about 30 neutrino events. The main result of the first experiment was to show that neutrinos from π decay were distinct from those produced in β decay as discussed in Sec. 6-2. Later experiments at Brookhaven and CERN used magnetic focusing for the pions in order to increase the neutrino flux, and also used more favorable geometries, finer-

Figure 6-11 Part of the spark-chamber array used in a second neutrino experiment at Brookhaven National Laboratory. There are a total of 184 six-foot-square aluminum plates in these chambers. (*Photograph courtesy of Professor L. M. Lederman, Columbia University.*)

Figure 6-12 Neutrino event observed in the chambers shown in Fig. 6-11. The neutrinos are incident from the left. (*Photograph courtesy of Professor L. M. Lederman, Columbia University.*)

grained detectors, and other improvements to increase the event rate and improve resolution.

THE TWO-COMPONENT NEUTRINO HYPOTHESIS

Experimental limits on the neutrino mass are $\lesssim 60$ eV for ν_e and $\lesssim 1.0$ MeV for ν_μ. If the ν is massless, then like the γ, it can only have two orientations of its spin vector, either parallel or antiparallel to its momentum. This is basically a property of the Lorentz transformation, but the following argument makes it plausible. If the ν had a component of spin normal to its momentum vector, this would mean that "parts" of it would move with $v > c$ (Fig. 6-13). Therefore the spin vector must be parallel or antiparallel to $\mathbf{v}$.

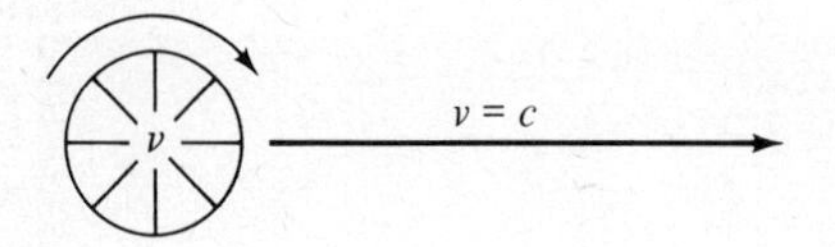

Figure 6-13 A neutrino traveling with velocity c cannot have a component of spin perpendicular to its velocity.

We define the *helicity* H of a beam of particles as the projection of the polarization vector $\mathbf{P}$ on the velocity vector $\mathbf{v}$,

$$H \equiv \frac{\mathbf{P} \cdot \mathbf{v}}{Pv} = \cos\theta \qquad (6\text{-}5)$$

where θ is the angle between the two vectors. Thus a helicity of $+1$ corresponds to the spin vector along $\mathbf{v}$ (right-handed), and -1 to the spin opposite to $\mathbf{v}$ (left-handed).

We might expect to find both right- and left-handed neutrinos in nature. In fact, it appears that only left-handed neutrinos and right-handed antineutrinos exist (Fig. 6-14). In this sense the weak interactions violate C and P invariance to the maximum possible extent. The two-component neutrino hypothesis is built into weak-interaction theory in the mathematical form chosen to describe the interaction. We shall not go into how this is done in detail, but an immediate consequence is that electrons and muons produced in weak decays have a helicity $H_{\pm} = \pm\beta$, where βc is the velocity of the lepton. This means that high-energy electrons from weak decays are almost completely longitudinally polarized with spin antiparallel to the momentum, and high-energy positrons are almost completely polarized with spin parallel to the momentum. This in turn can cause important correlations between spins in weak

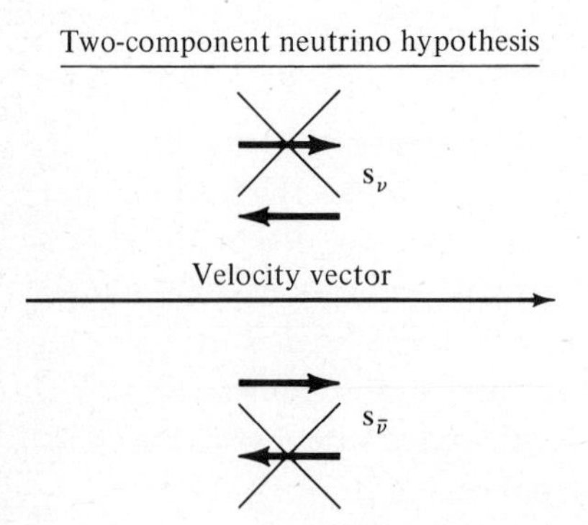

Figure 6-14 The two-component neutrino hypothesis postulates that there exist only neutrinos with spin antiparallel to their velocity vector and antineutrinos with spin parallel to the velocity.

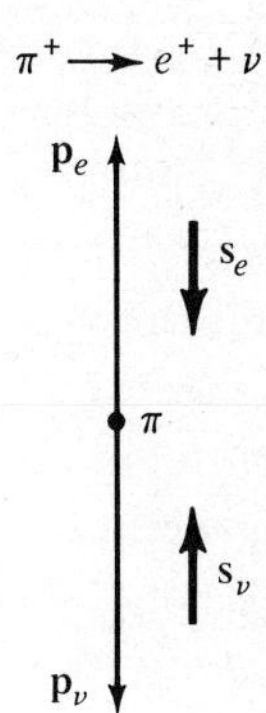

Figure 6-15 Momentum and spin vectors in the decay $\pi^+ \to e^+ + \nu$. The positrons must be emitted with the wrong helicity to conserve angular momentum.

decays because of angular-momentum considerations. For example, the decay $\pi \to e + \nu$ is greatly suppressed because the electrons must be emitted with the "wrong" spin direction (Fig. 6-15). This introduces a factor $(1 - \beta) \sim m_e^2/m_\pi^2$ in the decay rate.[1]

[1] For $\pi \to e + \nu$, this factor is $\approx 2 \times 10^{-5}$; for $\pi \to \mu + \nu$, it is ≈ 0.5.

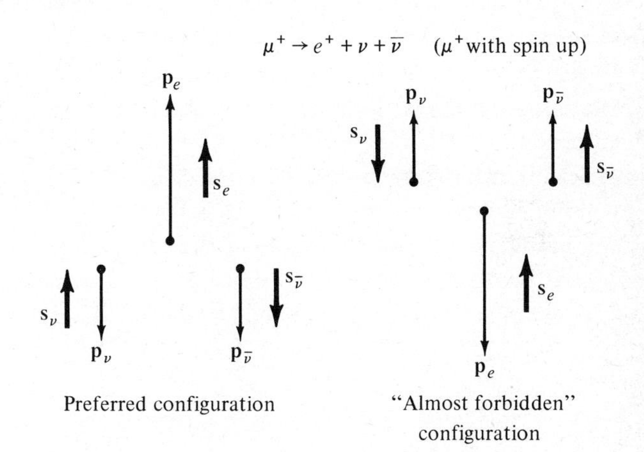

Figure 6-16 Momentum and spin vectors in the decay $\mu^+ \to e^+ + \nu + \bar{\nu}$. The parent muon is assumed to have spin up. The highest-energy positrons (in the muon rest system) are those which come off antiparallel to *both* neutrinos, as shown in these examples. The configuration at the right is strongly suppressed because the electron must come off with the wrong helicity. Because of this correlation the high-energy positrons tend to come off along the muon spin direction.

Another example is the correlation between the muon spin and decay-electron distribution discussed earlier. For $\mu^+ \rightarrow e^+ + \nu + \bar{\nu}$, the preferred and "almost forbidden" configurations are illustrated in Fig. 6-16. As might be expected, this correlation is strongest for electrons with the largest velocities.

A word of caution must be introduced here. Violations of C and P invariance are not necessarily associated with the presence of neutrinos. For example, the decay $\Lambda^0 \rightarrow p + \pi^0$ shows large P violations, as discussed previously, even though no neutrinos are involved.

6-6 INVARIANCE UNDER TIME REVERSAL T

Classically T invariance has a relatively simple interpretation. If we replace t with $-t$ in the equations of motion, we should obtain an equally acceptable solution. Newton's laws are obviously T invariant since they include only a second derivative with respect to time. Note that the time-reversed solution does not have to be an especially probable one. In fact in a large system of many particles, the time-reversed solution will often be a highly improbable situation.

In quantum mechanics the application of T invariance and its consequences are more subtle. It is hard to make generalizations, but an oversimplified statement of the consequences of T invariance is that, subject to the reservations given later, observables cannot depend on variables which are odd under T. Under the T operation,

$$t \rightarrow -t$$
$$\mathbf{r} \rightarrow +\mathbf{r}$$
$$\mathbf{v} \rightarrow -\mathbf{v}$$
$$\mathbf{p} \rightarrow -\mathbf{p}$$
$$\mathbf{s} \rightarrow -\mathbf{s}$$

As an example of a T-forbidden term, consider the decay $K \rightarrow \pi + \mu + \nu$. The decay muons are generally polarized, and the polarization can be measured by stopping the muons and observing the asymmetry of the electrons from their decay. T invariance forbids a component of polarization normal to the decay plane (specified by the vector product $\mathbf{p}_\pi \times \mathbf{p}_\mu$). Such a polarization, termed a *transverse* polarization, would indicate that the average of the term $\mathbf{s}_\mu \cdot (\mathbf{p}_\pi \times \mathbf{p}_\mu)$ is nonzero; in other words, the muon spin $\mathbf{s}_\mu$ would have a preferred direction relative to the vector $\mathbf{p}_\pi \times \mathbf{p}_\mu$. Such a term is odd under T and should average to zero if there is no preferred direction of time.

Experiments designed to look for a muon polarization normal to the decay plan in K^0 decay are consistent with no violation of T invariance, and place upper limits $\sim$1 percent for a T-violating component. Similar experiments have been carried out in β decay and Λ^0 decay with negative results. There is however evidence for a violation of T invariance in the weak decay $K^0 \rightarrow 2\pi$. This will be discussed in the next chapter.

As mentioned above, the nonexistence of terms which are odd under T is only expected under certain circumstances. The proof that terms such as $\mathbf{s}_\mu \cdot (\mathbf{p}_\pi \times \mathbf{p}_\mu)$ are forbidden by T invariance makes use of first-order perturbation theory. If perturbation theory is not applicable to a particular problem, such terms may appear. An example of a situation where such a term appears but is not T violating is in p-p elastic scattering. Here there generally *is* a polarization of the outgoing protons that is normal to the scattering plane and that corresponds to a term $\mathbf{s}_{p'} \cdot (\mathbf{p}_p \times \mathbf{p}_{p'})$. This is "allowed" because of the strong interaction between the two protons. In situations where there is an electromagnetic interaction such as in the decay $K^0 \rightarrow \pi^- + \mu^+ + \nu$ (where there is an electromagnetic final-state interaction), a small transverse polarization ($\sim\frac{1}{137}$) may appear.

It is interesting to point out the contrast between strong and weak interactions as far as allowed components of polarization are concerned. In the weak decays there are often large longitudinal components of polarization. Since such components correspond to terms like $\mathbf{s} \cdot \mathbf{p}$, they are P violating. Such terms do not occur in strong or electromagnetic interactions. On the other hand, in reactions having two or more *hadrons* in the final state, we can expect large transverse polarizations; otherwise T invariance forbids their appearance.

Another prediction of T invariance that is accessible to experimental test is the *detailed balance theorem.* This theorem relates cross sections for a reaction and its inverse:

$$\frac{\sigma(A \to B)}{\sigma(B \to A)} = \frac{p_b^2(2s_b + 1)(2s_B + 1)}{p_a^2(2s_a + 1)(2s_A + 1)} \tag{6-6}$$

where $s_a, \ldots$ are spins, and p_a and p_b are c.m.s. momenta, and it is understood that the c.m.s. energy is the same for both the forward and inverse reactions. The proof of detailed balance does not require perturbation theory, so that it is true for strong as well as electromagnetic and weak interactions. Tests of detailed balance have been made for strong interactions, for example, by comparing cross sections for $p + p \to \pi^+ + d$ with those for the inverse. Good agreement has been found. Tests of detailed balance in electromagnetic interactions are more difficult because of the smaller cross sections. Experiments to compare $n + p \to \gamma + d$ with its inverse have been carried out, and the results are consistent with T invariance.

Time-reversal violation is currently a subject of great interest because recent experiments have shown a violation of T invariance in the decay $K^0 \to 2\pi$ (Chap. 7). This is closely tied in with the question of invariance under the combined operation CP, which will be discussed further in the next chapter. The connection between T invariance and CP invariance comes about because of the so-called CPT theorem which we shall now discuss.

6-7 THE CPT THEOREM (LÜDERS–PAULI THEOREM)

An important theorem in quantum field theory, which can be proved under quite general assumptions, states that observables are invariant under the combined operation CPT (in any order). The assumptions required are Lorentz invariance and microscopic causality. (The latter essentially is the condition that signals cannot propagate faster than the velocity of light, even over microscopic distances.)

Aside from its theoretical basis, CPT invariance rests on a very firm experimental footing. Some of its predictions are:

1. The existence of an antiparticle for every particle
2. The equality of masses, lifetimes, magnetic moments, etc., of particles and antiparticles

The equality of lifetimes has been checked to an accuracy of about 0.1 to 1 percent for μ's, π's, and K's, and the equality of masses has been checked for many particles with varying degrees of accuracy. By a somewhat indirect argument (Sec. 7-2) we know that the K^0 and $\bar{K}^0$ masses are equal to about 1 part in 10^{14}. In addition, the magnetic moment anomaly $g - 2$ is known to be equal for μ^+ and μ^- to within about 0.1 percent.

In order that CPT invariance be preserved, if any one of C, P, or T is violated, one of the others must also be violated in a complementary way. Thus, for example, a T violation implies a CP violation and vice versa.

PROBLEMS

6-1 Estimate the probability of a neutrino's passing directly through the earth without interacting. Take the neutrino total cross section to be $\sim 10^{-14}$ of that for high-energy neutrons. (The radius of the earth is $\approx 6.4 \times 10^6$ m; its mass is $\approx 6 \times 10^{24}$ kg.) Repeat for a high-energy neutron. The earth is constantly being bombarded by a flux of neutrinos $\sim 4 \times 10^{10}$ cm^{-2}/sec from the sun. Make a rough estimate of the number of neutrino-induced reactions in our bodies per day.

6-2 List all the reactions associated with leg C of the tetrahedron, Fig. 6-3. Discuss briefly the practical significance, if any, of each.

6-3 Devise possible diagrams involving a four-fermion coupling contained in Fig. 6-3 for

$$\Lambda^0 \rightarrow p + \pi^-$$
$$K^+ \rightarrow \pi^0 + e^+ + \nu$$
$$\Xi^0 \rightarrow \Lambda^0 + \pi^0$$

6-4 Use the procedures discussed in Chap. 4 to calculate the ratio of decay rates for $K^+ \rightarrow e^+ + \nu$ and $K^+ \rightarrow \mu^+ + \nu$. Compare them with the observed rates in Table 1-1. Explain the discrepancy, and try to account for it quantitatively.

6-5 Devise a way of obtaining a beam of polarized muons.

6-6 Show that the existence of an electric dipole moment of the neutron is separately forbidden by P and T invariance. (The electric dipole moment **D** must lie along the neutron spin direction **s**, as does the magnetic dipole moment $\boldsymbol{\mu}$.)

6-7 Make a reasonable estimate of the mean energy of neutrinos produced by 15-GeV protons for a geometry like that in Fig. 6-10. (The relevant factors are the average pion energy, the pion-decay probabilities, and the energy of the neutrinos from the pion decay.)

6-8 What can be predicted about the polarization of muons from $\pi \rightarrow \mu + \nu$ (in the rest system of the pion)? What correlation between the electron and neutrino momenta would arise in the decay $K^0 \rightarrow \pi^- + e^+ + \nu$ because of helicity considerations?

6-9 Which of the following reactions can proceed by a first-order weak interaction?

(*a*) $\mu^+ \rightarrow e^+ + e^+ + e^-$

(*b*) $\mu^- + p \rightarrow e^- + p$

(*c*) $\pi^- \rightarrow \mu^- + \bar{\nu}_\mu + \nu_e + \bar{\nu}_e$

(*d*) $\Omega^- \rightarrow \Lambda^0 + \pi^-$

(*e*) $K^+ \rightarrow \pi^+ + \pi^+ + e^- + \bar{\nu}_e$

6-10 Consider in some detail an experiment to test T invariance in electromagnetic or weak interactions. In particular, discuss:

(*a*) The principle of the experiment

(*b*) The experimental technique

(*c*) Systematic and other errors, and their effect on the interpretation of the data

(*d*) The feasibility of the experiment with current techniques

CHAPTER 7

the K^0-$\bar{K}^0$ system: CP violation

As mentioned in the previous chapter, a small violation of both CP and T invariance has been found in K^0 decays. To understand the evidence for this violation, we must first discuss the properties of the K^0-$\bar{K}^0$ system. As we shall see, this is a unique and fascinating system which exhibits striking phenomena of a purely quantum-mechanical nature.

7-1 THE K^0–$\bar{K}^0$ SYSTEM: INTRODUCTION

K^0's (and $\bar{K}^0$'s) are observed to decay with two distinct lifetimes, as shown schematically in Fig. 7-1. The shorter-lived component $K_S{}^0$ has a mean life of approximately 10^{-10} sec, and the longer-lived $K_L{}^0$ has a mean life of $\approx 5.2 \times 10^{-8}$ sec. The decay modes of the two components are quite different; the short-lived component decays primarily into two π's, while the long-lived one decays primarily into three π's or into a π and a lepton pair (Table 1-1). Because of the large disparity in lifetimes, a K^0 beam after a time of about 10^{-9} sec will be almost completely $K_L{}^0$.

For definiteness, let us consider a beam of K^0's initially produced in the reaction

$$\pi^- + p \to K^0 + \Lambda^0$$

The K^0's are produced in a strong reaction and therefore have a well-defined strangeness of $+1$. However they decay through a weak interaction, which does not conserve strangeness. Furthermore the only quantum number which distinguishes a K^0 and $\bar{K}^0$ is strangeness. K^0's can therefore be transformed into $\bar{K}^0$'s through virtual

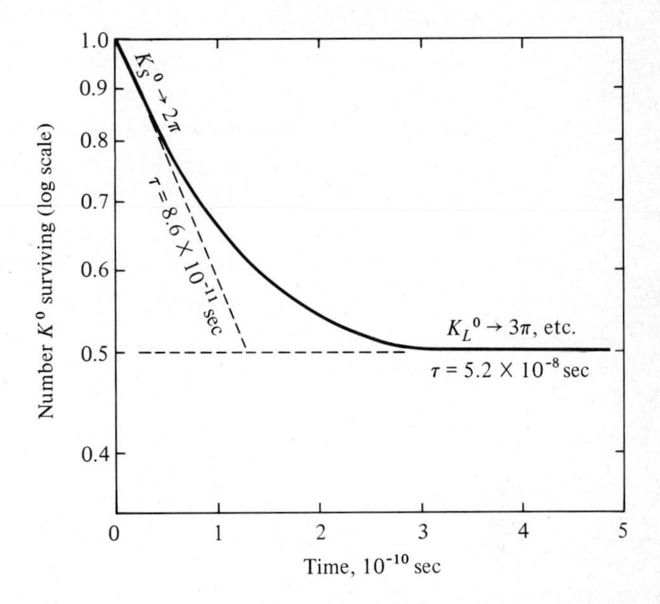

Figure 7-1 Number of K^0 surviving versus time after production. Two distinct lifetimes are observed, with the shorter-lived component decaying principally into two pions and the long-lived component decaying into three pions or a pion plus leptons (see Table 1-1).

second-order weak interactions such as that shown in Fig. 7-2.[1] We are interested in describing the time dependence of a system which at $t = 0$ is pure K^0. Quantum mechanically the K^0 and $\bar{K}^0$ form a degenerate system. (The masses or energies are equal.) The

[1] Note that this is the only example of a particle and antiparticle which are distinct yet coupled, since this requires nonzero strangeness, but zero charge, lepton number, and baryon number.

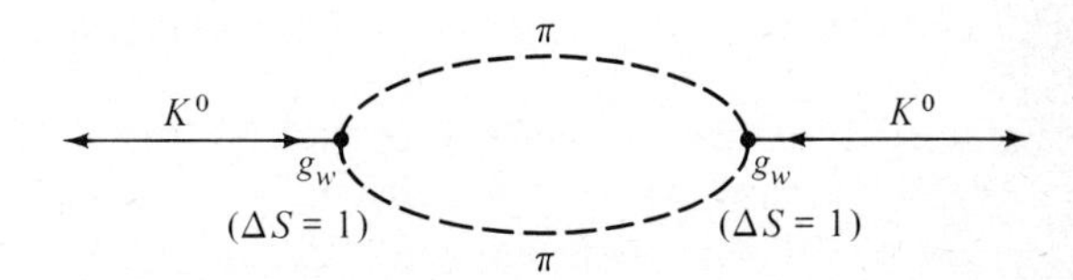

Figure 7-2 Coupling of K^0 and $\bar{K}^0$ through second-order weak interactions. Both the K^0 and $\bar{K}^0$ can virtually dissociate into two pions through a weak interaction.

decay through weak interactions can be considered a perturbation which breaks the degeneracy. The first problem is then to find the appropriate linear combination of eigenfunctions of the unperturbed system (i.e., of K^0 and $\bar{K}^0$) to describe the perturbed system.

For the moment, we neglect the small CP violation. Then in a weak decay the CP quantum number of a state is conserved. The K^0 (or $\bar{K}^0$) is not an eigenstate of CP, since $\mathrm{C}(K^0) = \bar{K}^0$ and P does not affect the wave function (except for a possible change of sign which can be disregarded).[1] The desired linear combinations of K^0 and $\bar{K}^0$ are therefore the eigenfunctions of CP. These can be readily constructed as follows:

$$K_1{}^0 = \frac{1}{\sqrt{2}}(K^0 + \bar{K}^0)$$
$$K_2{}^0 = \frac{1}{\sqrt{2}}(K^0 - \bar{K}^0) \tag{7-1}$$

Now $\mathrm{CP}(K_1{}^0) = +K_1{}^0$, and $\mathrm{CP}(K_2{}^0) = -K_2{}^0$, so these are evidently eigenstates of CP with eigenvalues $+1$ and -1 respectively.

We can identify $K_1{}^0$ and $K_2{}^0$ with the physical states $K_S{}^0$ and $K_L{}^0$ (neglecting CP violation) by the following argument: The $K_S{}^0$ has the prominent decay mode $K_S{}^0 \rightarrow 2\pi^0$, and the two-$\pi^0$ system is clearly even under CP since $\mathrm{C}(\pi^0) = \pi^0$ and two π^0's must be in a state with even orbital parity. The short-lived component is therefore $K_1{}^0$, and the long-lived is $K_2{}^0$. Since the phase space available for the two-pion decay is much larger than that for the three-pion decay, the $K_1{}^0$ lifetime is much shorter than that of the $K_2{}^0$.

We can invert Eqs. (7-1) to obtain

$$K^0 = \frac{1}{\sqrt{2}}(K_1{}^0 + K_2{}^0)$$
$$\bar{K}^0 = \frac{1}{\sqrt{2}}(K_1{}^0 - K_2{}^0) \tag{7-2}$$

Thus a beam of K^0's formed through a strong interaction is initially an equal mixture of $K_1{}^0$ and $K_2{}^0$.

The K^0 and $\bar{K}^0$ are particle and antiparticle, so their masses must be exactly equal by the CPT theorem. The $K_1{}^0$ and $K_2{}^0$, however,

[1] For simplicity we use the particle symbol to represent the wave function.

can and do have different masses. As discussed earlier (Sec. 5-3), the masses of the particles are thought to be generated by their interactions. The largest contribution to the hadron masses is from the strong interactions. Differences in electromagnetic interactions between, say, the K^+ and K^0 cause a mass difference between these particles of order $m_K/137$. In the same spirit the difference in the weak interactions of the K_1^0 and K_2^0 arising from their different CP quantum numbers leads to a mass difference of order $10^{-14}\, m_K$, or about 5×10^{-6} eV. Despite its extremely small size this mass difference has been measured quite accurately and is found to be $\approx 3.55 \times 10^{-6}$ eV, with the long-lived K^0's being slightly more massive (Sec. 7-5).

7-2 EVOLUTION OF A K^0 BEAM WITH TIME

If we generate a K^0 beam by a strong interaction, the wave function of the system at $t = 0$ from the first of Eqs. (7-2) will have the form

$$\begin{aligned} \psi(0) &= K^0 \\ &= \frac{1}{\sqrt{2}} (K_1^0 + K_2^0) \end{aligned}$$

As time progresses, the K_1^0 and K_2^0 amplitudes decay with their characteristic lifetimes. The *intensity* of K_1^0 or K_2^0 components can be obtained by squaring the appropriate coefficient in $\psi(t)$. The *amplitudes* therefore contain a factor $e^{-t/2\tau}$. They also contain a factor $e^{-iEt/\hbar}$ which describes the time dependence of an energy eigenstate in quantum mechanics.[1] In the rest system of the K^0 the latter factor can be written e^{-imt}, where m is the mass and we adopt units such that $\hbar = c = 1$. The complete wave function for the system can therefore be written

$$\begin{aligned} \psi(t) &= \frac{1}{\sqrt{2}} (K_1^0 e^{-t(1/2\tau_1 + im_1)} + K_2^0 e^{-t(1/2\tau_2 + im_2)}) \\ &= \frac{1}{\sqrt{2}} e^{-im_1 t} (K_1^0 e^{-t/2\tau_1} + K_2^0 e^{i\,\Delta m\, t}) \end{aligned} \tag{7-3}$$

[1] Usually this factor drops out when absolute squares are taken, so it is neglected in problems where the potential function describing the interaction is time independent, but in this case it is very important.

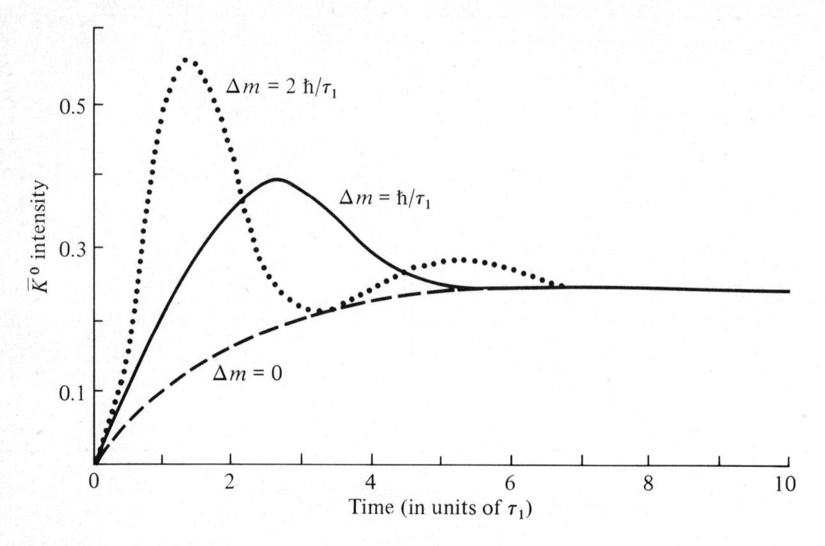

Figure 7-3 The variation of the intensity of the $\bar{K}^0$ component of a beam, initially pure K^0, for different values of Δm.

where $\Delta m \equiv m_1 - m_2$ and we neglect the slowly varying factor $e^{-t/2\tau_2}$ for simplicity ($\tau_2 \approx 600\tau_1$).

Reexpressing K_1 and K_2 in terms of K^0 and $\bar{K}^0$, we find

$$\begin{aligned}\psi(t) &= \tfrac{1}{2}e^{-im_1t}[e^{-t/2\tau_1}(K^0 + \bar{K}^0) + e^{i\,\Delta m\,t}(K^0 - \bar{K}^0)] \\ &= \tfrac{1}{2}e^{-im_1t}[K^0(e^{-t/2\tau_1} + e^{i\,\Delta m\,t}) + \bar{K}^0(e^{-t/2\tau_1} - e^{i\,\Delta m\,t})]\end{aligned}$$

The intensity of the K^0 component is obtained by taking the absolute square of the coefficient of K^0,

$$I(K^0) = \tfrac{1}{4}[e^{-t/\tau_1} + 1 + 2e^{-t/2\tau_1}\cos(\Delta m\,t)] \tag{7-4}$$

Similarly

$$I(\bar{K}^0) = \tfrac{1}{4}[e^{-t/\tau_1} + 1 - 2e^{-t/2\tau_1}\cos(\Delta m\,t)] \tag{7-5}$$

The time variation of the $\bar{K}^0$ component is plotted in Fig. 7-3 for several values of Δm.[1] The most interesting feature of the curves

[1] Note that if the K^0 and $\bar{K}^0$ masses were different, it would drastically affect the expected time dependence. The general agreement with theory can therefore be considered evidence that

$$\begin{aligned}|m_{K^0} - m_{\bar{K}^0}| &\ll \Delta m \\ &\approx 3.6 \times 10^{-6}\ \text{eV}\end{aligned}$$

is their oscillatory behavior caused by the interference of the $K_1{}^0$ and $K_2{}^0$ amplitudes with their slightly different time dependence. Note that this oscillation is on a *macroscopic* scale (~10-cm wavelength for typical K^0 beams).

7-3 CP VIOLATION

The discussion so far has assumed that CP is conserved in K^0 decays. From the preceding we should expect that after a K^0 beam has gone for, say, 10 mean lifetimes of the $K_1{}^0$ component, it would be an essentially pure $K_2{}^0$ beam and no $K^0 \rightarrow \pi^+ + \pi^-$ decays would be observed. This in fact is not the case, as first shown in a famous experiment by Christenson, Cronin, Fitch, and Turlay.[1] They showed that even after a K^0 beam has traveled many mean lifetimes for the $K_1{}^0$ component, about 0.3 percent of the K^0's still decay into two pions though, as we have discussed earlier, the two-π system is a CP even state, while the $K_2{}^0$ is CP odd.

The experimental arrangement used by Christenson et al. was quite straightforward and is shown in Fig. 7-4. A neutral beam is formed by sweeping out charged particles with bending magnets. The principal component of the beam is neutrons, but these do not decay in significant numbers. The experiment consisted in looking for pairs of charged pions coming out of the decay region, which is many $K_1{}^0$ mean lifetimes away from the K^0 production target. If the K^0 decays into just two pions, the pion tracks will be coplanar with the incident beam, and the mass of the parent particle, as reconstructed from the vector momenta of the pions, will equal the K^0 mass. The experiment showed a clean signal of two-pion decays, which indicated that about 0.3 percent of the K^0 beam decayed into two charged pions. Further experiments have since plugged any theoretical or experimental loopholes, and the conclusion that CP is violated, at least to a small extent in K^0 decays, now seems inescapable.

As a result of this CP violation, the physical states $K_L{}^0$ and $K_S{}^0$ are not pure eigenstates of CP but linear combinations of $K_1{}^0$

[1] J. H. Christenson, J. W. Cronin, V. L. Fitch, and R. Turlay, *Phys. Rev. Letters* **13,** 138 (1964).

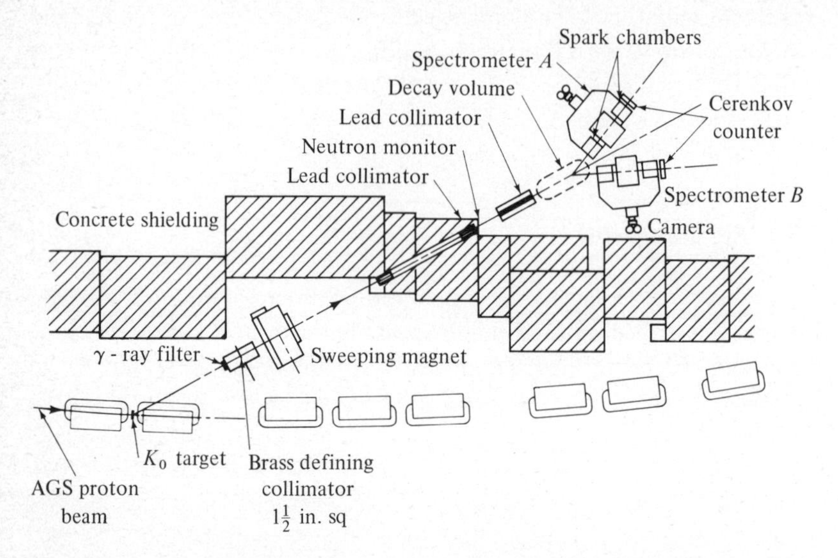

Figure 7-4 The experimental arrangement used by Christenson et al. A K^0 beam is formed when the proton beam strikes a target in the AGS. Some of the long-lived K^0's decay in the decay volume. The two magnetic spectrometers measure the momentum and angles of charged particles coming from the decay volume. [*From J. H. Christenson et al., Phys. Rev.* **140**, B74 (1965).]

and K_2^0, the eigenstates defined above. Since the violation is quite small, $K_S^0 \approx K_1^0$ and $K_L^0 \approx K_2^0$. The effect of the CP-violating interaction is to continuously change the K_2^0 component in the "stale" K^0 beam into K_1^0, which decays by $K_1^0 \rightarrow 2\pi$ (Fig. 7-5). We really know little about the CP-violating interaction in the box, other than that it connects states having CP $= +1$ with those having CP $= -1$. A measure of the relative strength of this

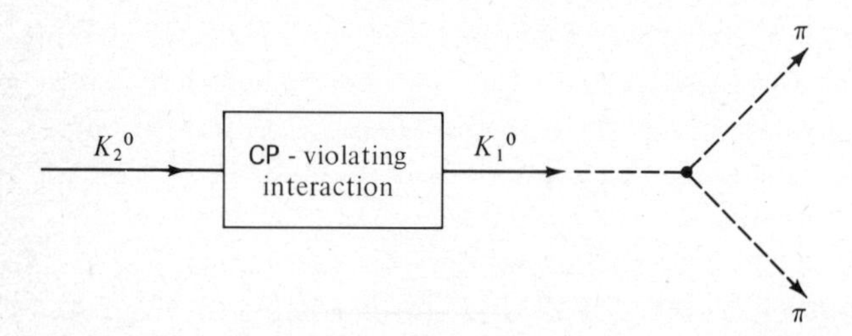

Figure 7-5 The CP-violating interaction continuously changes K_2^0's with CP quantum number -1 to K_1^0's with CP $= +1$.

interaction is the ratio of the amplitude of the forbidden decay $K_L{}^0 \to \pi^+ + \pi^-$ to that of the allowed decay $K_S{}^0 \to \pi^+ + \pi^-$,

$$\begin{aligned} |\eta_{+-}| &\equiv \left| \frac{A(K_L{}^0 \to \pi^+ + \pi^-)}{A(K_S{}^0 \to \pi^+ + \pi^-)} \right| \\ &= \left[\frac{\text{rate}(K_L{}^0 \to \pi^+ + \pi^-)}{\text{rate}(K_S{}^0 \to \pi^+ + \pi^-)} \right]^{\frac{1}{2}} \\ &\approx 1.95 \times 10^{-3} \end{aligned} \tag{7-6}$$

Unlike the large C and P violations in weak interactions, the CP violation appears to be a small, almost negligible effect. In fact no other manifestation of CP violation outside of the K^0-$\bar{K}^0$ system has yet been positively identified. It is natural to attribute the CP violation to the weak interaction. However we must remember that the box in Fig. 7-5 can contain virtual strong and electromagnetic interactions. Bernstein, Feinberg, and Lee, noting that $|\eta_{+-}| \sim \alpha = \frac{1}{137}$, have suggested that the violation occurs in the electromagnetic interactions between hadrons and appears in the K^0 system because of diagrams such as Fig. 7-6.[1] This hypothesis, aside from explaining in a natural way the smallness of the effect, has the virtue of predicting observable effects in other reactions (see Sec. 7-4).

Another possibility is that the violation is due to a new interac-

[1] J. Bernstein, G. Feinberg, and T. D. Lee, *Phys. Rev.* **139,** B1650 (1965). The authors pointed out that CP and C invariance had not been well tested in electromagnetic processes involving hadrons.

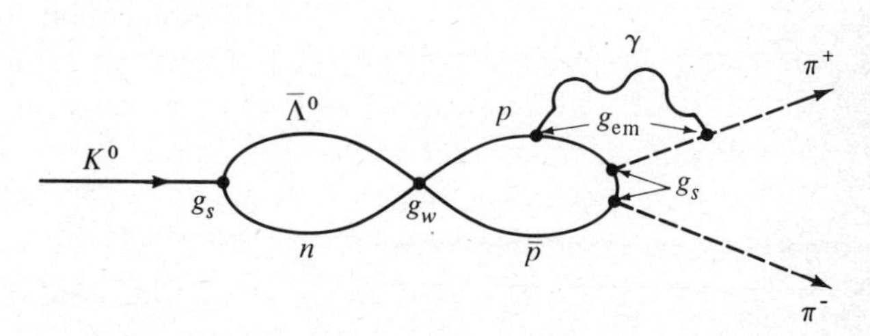

Figure 7-6 A second-order diagram for K^0 decay involving an electromagnetic interaction between hadrons. Bernstein, Feinberg, and Lee propose that the CP violation occurs at the electromagnetic vertices. Diagrams such as this interfering with the main diagram (without the photon exchange) would give a $K_L{}^0 \to 2\pi$ rate on the order of that observed.

tion. L. Wolfenstein has proposed that there may exist a CP-violating "superweak" interaction which couples K^0 and $\bar{K}^0$ directly (so that unlike the weak interaction, it violates strangeness by two units).[1] This interaction competes with the second-order weak interaction shown in Fig. 7-2, and its strength need only be $\sim 10^{-2}$ of that of the second-order weak interaction to account for the observed effect in K^0 decays. It turns out that the effects of such an interaction outside of the K^0 system are far too small to show up experimentally.

The superweak and certain other proposed theories predict that

$$\eta_{00} = \eta_{+-}$$

where

$$\eta_{00} \equiv \frac{A(K_L{}^0 \to \pi^0 + \pi^0)}{A(K_S{}^0 \to \pi^0 + \pi^0)} \tag{7-7}$$

Unfortunately the measurement of the $K_L{}^0 \to 2\pi^0$ rate is an extremely difficult experiment because it involves all neutral particles. Presently available data show $\eta_{00} \approx \eta_{+-}$. This is consistent with the superweak theory, but unfortunately is also consistent with most other theories.

7-4 TESTS OF CP, C, AND T

Unless further manifestations of CP violation are found, it will be very difficult to decide between the various possible theories. There is currently a great deal of experimental effort in this direction. It turns out that possible tests of CP invariance are quite restricted; the search for $K_L{}^0 \to 2\pi$ is one of the very few sensitive tests. If combined CPT invariance is valid, the CP violation implies a complementary violation of T invariance. Tests of T invariance were discussed briefly in Sec. 6-6. No compelling evidence for a T violation has yet been found outside the K^0 system.[2]

Theories which attribute the CP violation to the strong or electromagnetic interactions of hadrons would also predict violations of C invariance in strong or electromagnetic interactions. This is

[1] L. Wolfenstein, *Phys. Rev. Letters* **13,** 562 (1964).

[2] It can be shown that the available data on $K^0 \to 2\pi$ decays show direct evidence of a T violation, without recourse to the CPT theorem. See M. Gourdin, *Nucl. Phys.* **B3,** 207 (1967); R. Casella, *Phys. Rev. Letters* **22,** 554 (1969).

because P invariance in these interactions has been tested with a high degree of accuracy, notably in the absence of circular polarization of certain nuclear γ rays.[1] A CP violation therefore implies a complementary violation of C, so that P invariance is preserved. Tests of C invariance in strong and electromagnetic interactions have consequently become of interest. C invariance, for example, predicts the equality of masses of particle and antiparticle, but this is also guaranteed independently by the presumably stronger requirement of CPT invariance. Useful tests of symmetry under charge conjugation independent of CPT can only be carried out in systems which are self-conjugate under C. These include π^0, η^0, and all particle-antiparticle systems. The most sensitive tests to date have been made in the decays

$$\eta^0 \underset{\text{em}}{\rightarrow} \pi^+ + \pi^- + \pi^0$$

and

$$\eta^0 \underset{\text{em}}{\rightarrow} \pi^+ + \pi^- + \gamma$$

C invariance requires that the momentum distributions of the π^+ and π^- (relative to the π^0 or γ) be symmetrical. Present data are only marginally consistent with C invariance, but the experiments are difficult and past experience indicates it is wise to take a conservative attitude toward such discrepancies.

Another important test of these invariance principles concerns the possible existence of electric dipole moments of the elementary particles,[2] in particular of the neutron, for which the most sensitive measurements have been made. The existence of an electric dipole moment is independently forbidden by symmetry under P, CP, and T (Prob. 6-6). The experimental upper limit for the neutron electric dipole moment is $\sim 2 \times 10^{-23}$ e-cm, where e is the electron charge.[3] Since the "size" of the neutron is $\sim 10^{-13}$ cm, a "large" electric dipole moment would be 10^{-13} to 10^{-14} e-cm. However the weak interactions must be invoked to provide the P violation, and this introduces a factor of $g_w^{\frac{1}{2}}$, or about 10^{-7}. Theoretical

[1] Such a polarization has been observed, but at a level $\sim 10^{-6}$, which is consistent with that expected from weak interactions.

[2] This could be considered to indicate that there is an asymmetry of the charge distribution along the direction of the magnetic moment.

[3] P. Miller, W. Dress, J. Band, and N. Ramsey, private communication. See also W. Dress et al., *Phys. Rev.* **170,** 1200 (1968).

estimates of the expected neutron electric dipole moment if the CP violation is in the electromagnetic interactions are on the order of 10^{-21} e-cm; but because of the uncertainty of the theoretical estimates, the fact that the experimental limit is considerably smaller cannot yet be regarded as strong evidence against an electromagnetic violation of CP.

Since there is at present no strong evidence for CP or T violations outside the K^0-$\bar{K}^0$ system, there is very little to go on in choosing between the many possible theories. As time goes on, if no further manifestations arise and if η_{00} turns out to be equal to η_{+-} within small errors, the weight of evidence will swing toward the superweak theory. Unfortunately the practical consequences of this theory are nill, and if the superweak theory turns out to be correct, it would seem almost as though nature had played a cosmic practical joke on physicists.

7-5 REGENERATION EXPERIMENTS

As discussed earlier, a $K_L{}^0$ beam is an approximately equal mixture of K^0 and $\bar{K}^0$ components. When a $K_L{}^0$ beam passes through matter, the strong interactions preferentially absorb the $\bar{K}^0$ component since the total cross sections for $\bar{K}^0$ on nucleons are typically

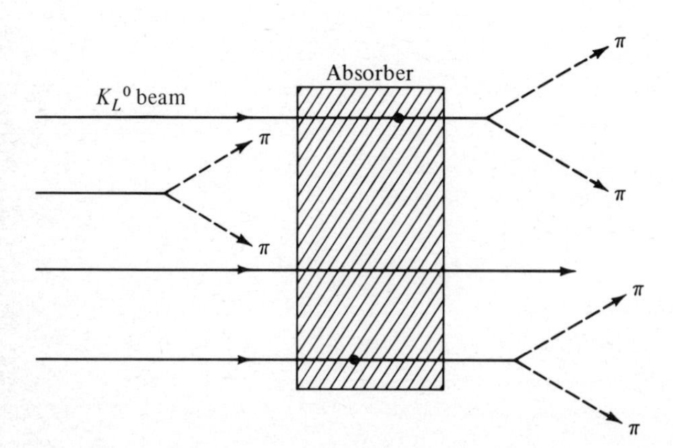

Figure 7-7 When a K_L beam passes through an absorber, some of the K_S component is regenerated. Downstream from the absorber, $K_S{}^0 \rightarrow 2\pi$ decays are observed.

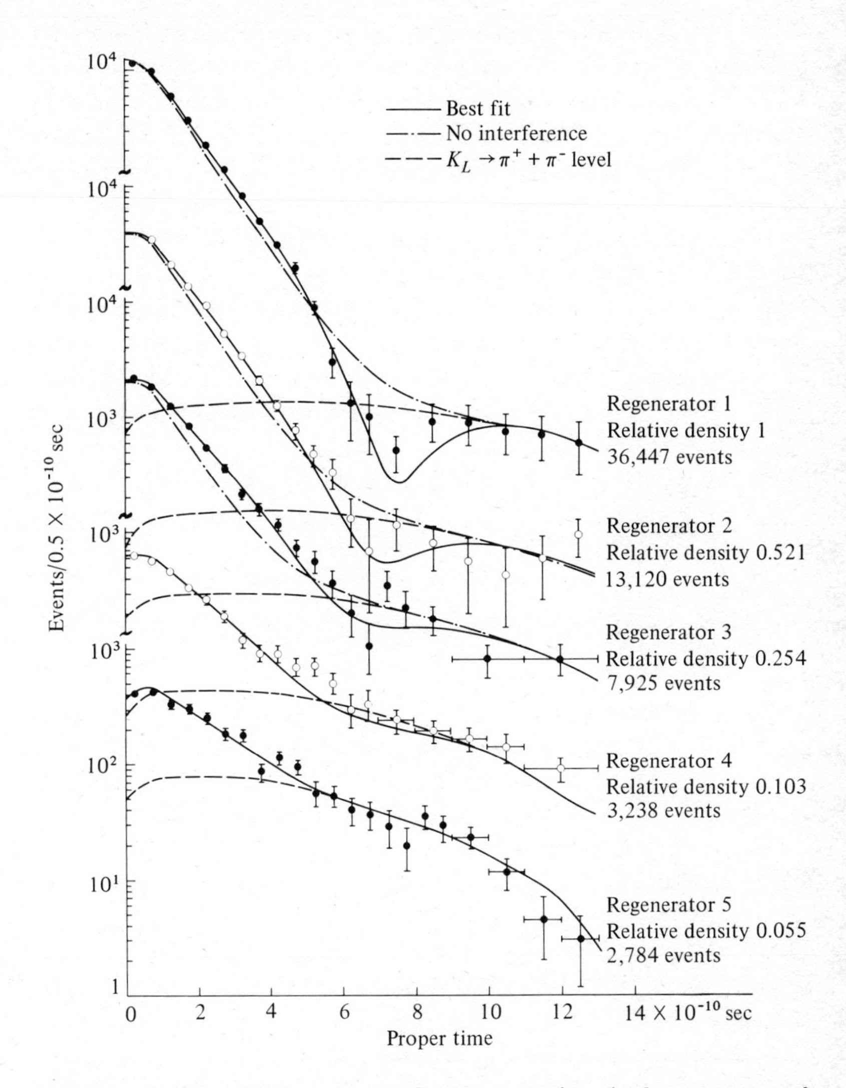

Figure 7-8 Number of $K^0 \to \pi^+ + \pi^-$ decays versus time (in the rest system of the K^0) observed with various regenerators. The solid curves are fits to the data, assuming interference between $K_L{}^0 \to 2\pi$ and $K_S{}^0 \to 2\pi$ with $\Delta m = 3.65 \times 10^{-6}$ eV. [*Data from H. Faissner et al., Phys. Letters* **B30**, 204 (1969).]

considerably larger than those for K^0.[1] This disturbs the K^0-$\bar{K}^0$ mixture and has the effect of renewing, or "regenerating," some of the $K_S{}^0$ component. Just downstream from the absorber the two-pion decays characteristic of $K_S{}^0$ will be observed (Fig. 7-7). However the total amplitude for two-pion decays will be the sum of the regenerated $K_S{}^0$ amplitude and that associated with the CP-violating two-pion decay of the $K_L{}^0$. The rate of two-pion decays versus time will be given by an equation similar to Eq. (7-4) with an interference term containing $\cos(\Delta m\, t)$. Experiments like that shown schematically in Fig. 7-7 have been carried out to measure Δm and the phase of η_{+-} (which is a complex parameter). In Fig. 7-8, data from one such experiment show the variation of the rate of $K^0 \to 2\pi$ events behind the regenerator with proper time. From this experiment and variations of it, Δm is now known with very high accuracy,

$$\begin{aligned}\Delta m &= m(K_S{}^0) - m(K_L{}^0)\\ &= -(3.553 \pm 0.0022) \times 10^{-6}\ \text{eV}\\ &\approx -0.464\,\frac{\hbar}{\tau_S}\end{aligned}$$

PROBLEMS

7-1 Does the fact that a K^0 can decay as either a $K_S{}^0$ or a $K_L{}^0$ with slightly different masses violate energy conservation? Could such a violation be detected?

7-2 Calculate $I(K^0)$ and $I(\bar{K}^0)$ for $\Delta m = 0.5\hbar/\tau_S$ from Eqs. (7-4) and (7-5).

7-3 Discuss some of the experimental and theoretical loopholes that might have "saved" CP invariance after the publication of the paper by Christenson et al. (Sec. 7-3). (A good starting point would be a search of *Physical Review Letters* and *Physics Letters* for relevant articles in the 12 months following publication of their article.)

7-4 In the experiment of Christenson et al., the mean K^0 momentum was 1.1 GeV/c. The distance from the K^0 production target to the decay region was 18.5 m. Estimate the ratio of the intensities of the $K_S{}^0$ and $K_L{}^0$ components at the decay region.

[1] The reason is that reactions such as $\bar{K}^0 + p \to \Lambda^0 + \pi^+$ are possible, but because of conservation of strangeness, $K^0 + p \to \Lambda^0 + \pi^+$ is not possible.

CHAPTER 8

the eightfold way, quarks, and SU(3) symmetry

Now this, O monks, is the noble truth of the way that leads to the cessation of pain: this is the noble Eightfold Way: namely, right views, right intention, right speech, right action, right living, right effort, right mindfulness, right concentration.

From the "Sutta of Turning the Wheel of the Doctrine"
by Buddha, Canon, *Samyutta v*, 420.

In this chapter we discuss the application of group-theoretical techniques to the strong interactions. To put things in their proper perspective, it is important to realize that not even a first approximation of a comprehensive theory of strong interactions exists. We are therefore willing to settle for a much less ambitious goal, to push general symmetry arguments as far as possible. Success in this undertaking to date has been real, though limited.

The approach is to postulate that the strong interactions have certain symmetries in addition to symmetry under rotations in isospin space. Just as the electromagnetic interaction partially breaks this symmetry under isospin rotation, we suppose these additional symmetries are broken by some part of the *strong* interaction (at the 10 percent level). As we shall see, this approach is closely tied to the question whether all particles so far discovered can be built up out of elementary particles called *quarks*.

Perhaps the most important goal of any theory is to classify the almost bewildering array of particles and resonances we discussed previously, just as Mendeleev did for the elements in the periodic table. The most successful attempt to date goes by the name given to it by one of its originators, Murray Gell-Mann, who called it *the eightfold way*.[1] The essential idea of this scheme is that the particles can be classified into multiplets whose structure and multiplicity are appropriate to the symmetry group called *the special unitary group of dimension* 3 [SU(3)] in group theory. Group theory is an abstract branch of mathematics which deals with the properties of algebraic systems under transformations which maintain certain symmetries of the algebra.[2] It is not our purpose here to introduce

[1] From the exhortation by Buddha quoted at the beginning of the chapter.

[2] In Appendix C we briefly summarize some of the terminology of group theory.

the reader to group theory. Rather, we shall proceed by analogy with isotopic spin, whose algebra (i.e., rules of combination) is appropriate to the symmetry group SU(2).

8-1 ISOTOPIC SPIN AND THE GROUP SU(2)

As we have discussed earlier, the strong interactions appear to depend only on the magnitude of the isospin vector $|\mathbf{T}|$, not on its components T_1, T_2, T_3. In other words, a system of strongly interacting particles is symmetrical with respect to rotations in isospin space. It is tempting to speculate on the physical significance of isospin space. However such an approach is not consistent with the spirit in which it was introduced.

Basically when we use the concept of isospin to classify a particle system, we are merely saying that to describe certain aspects of the behavior of the system, we can assign to it a quantity $\mathbf{T}$ whose algebra is just like that of an angular-momentum vector (in quantum mechanics). A group theorist would say such a system had a symmetry belonging to the special unitary group of dimension 2 [SU(2)].

As discussed earlier, particles can be classified as charge or isospin multiplets. Each multiplet (nucleon, Λ, Σ, . . .) has associated with it a quantum number T such that the multiplicity of the multiplet is $2T + 1$. Each particle in a multiplet is distinguished by a *single* quantum number T_3, possible values of which are $\pm T$, $\pm(T - 1)$, ... Furthermore, T_3 is an *additive* quantum number (since it is directly related to the charge by $T_3 = Q/e - B/2 - S/2$), that is, two particles with T_3 values T_3' and T_3'' give a state with $T_3 = T_3' + T_3''$. When two particles interact, their isospins can combine to give states with different multiplicities. For example, by combining a pion and nucleon (an isotopic triplet and a doublet) we can get states with $T = \frac{3}{2}$ or $T = \frac{1}{2}$ (a quartet and a doublet). We shall denote this by the symbolic equation.

$$\mathbf{3} \otimes \mathbf{2} = \mathbf{4} \oplus \mathbf{2}$$

A group theorist would recognize the above rules as those appropriate to the group SU(2). This group contains all transformations

which can be expressed by 2×2 matrices whose determinant is unity (*special*) and which are unitary. The allowed multiplicities 1, 2, 3, . . . , corresponding to particles of isospin 0, $\frac{1}{2}$, 1, . . . , are to be identified with the dimensions of the *irreducible representations* of the group SU(2).

As a model of a physical system whose algebra is contained in SU(2), let us consider systems built up out of protons ($T_3 = \frac{1}{2}$) and neutrons ($T_3 = -\frac{1}{2}$) and determine possible values of T and T_3 for such systems. With the rules for combining isospins given above, the answer is clear: the possible values are $T = 0, \frac{1}{2}, 1, \ldots$ and $T_3 = \pm T, \pm(T - 1), \ldots, 0$. Thus we have "predicted" the allowed values of T and T_3 for nuclei. This prediction might have been arrived at even if nucleons, as such, did not have an independent existence outside of the nucleus. One approach might have been to guess that the rules were those appropriate to the algebra of SU(2). Equivalently one might guess that the algebra was the same as that for combining ordinary spin in quantum mechanics.

The group-theoretical approach is the more general and powerful. However since it is not our purpose to teach group theory, we shall adopt a model approach to higher symmetries. A model for SU(3) analogous to that discussed above for SU(2) is to consider systems built up out of three "building blocks" [instead of two, the proton and neutron, as in SU(2)]. Gell-Mann coined the name *quark*[1]

[1] From a line in James Joyce's "Finnegan's Wake" which begins, "Three quarks for Muster Mark. . . ."

Murray Gell-Mann. Born 1929 in New York City. He obtained his Ph.D. from Massachusetts Institute of Technology at the age of 22. In 1953 Gell-Mann, and independently T. Nakano and K. Nishijima, proposed the existence of a new quantum number strangeness that was conserved in strong interactions but not in weak (this explained why K mesons and hyperons could be produced in strong interactions but decayed with a lifetime characteristic of a weak interaction). Gell-Mann and Y. Ne'eman independently developed a classification scheme based on SU(3) symmetry for the hadrons. Gell-Mann and G. Zweig subsequently postulated the existence of three fractionally charged particles, to which Gell-Mann gave the name *quarks*, as the fundamental triplet which served as the basis for the SU(3) symmetry of the baryons. In 1969 Gell-Mann received the Nobel Prize "for his contributions and discoveries concerning the classification of elementary particles and their interactions." (*From the Meggers Gallery of Nobel Laureates, American Institute of Physics.*)

for members of a hypothetical elementary triplet out of which the elementary particles can be "constructed."

8-2 SUPERMULTIPLETS AND SU(3)

We shall return to this model of SU(3) in the next section. First we need to discuss some preliminaries. The particles are already grouped in charge or isopin multiplets with SU(2) symmetry. We want to extend this classification to form *supermultiplets* which are representations of a higher symmetry. Why pick SU(3) when there are many other possible schemes? The only answer to this (as for most similar questions in physical theory) is, "It works."

In order to identify a particle in a given supermultiplet, we have to specify not only T_3, but at least one other quantum number. This could be strangeness, but it turns out to be more convenient to use the *hypercharge* Y, where

$$Y \equiv S + B$$

with B as the baryon number.

Inspection of Table 1-1 shows that there are eight spin $\frac{1}{2}^+$ baryons, N^+, N^0, Λ^0, Σ^+, Σ^-, Σ^0, Ξ^-, and Ξ^0. It is convenient to exhibit these on a plot of Y versus T_3, as shown in Fig. 8-1, where the scales are chosen to emphasize the symmetry. A group theorist

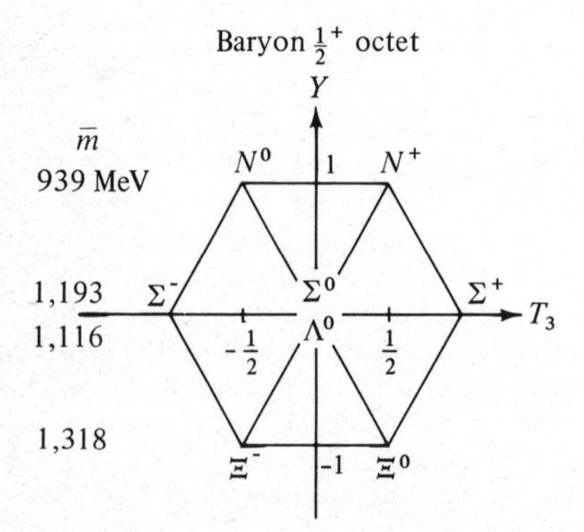

Figure 8-1 Plot of hypercharge versus T_3 for the spin $\frac{1}{2}^+$ baryons. The scales have been chosen to emphasize the symmetry.

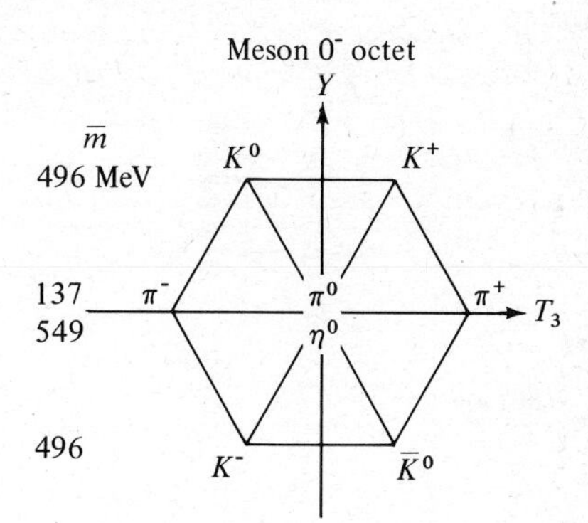

Figure 8-2 The meson 0⁻ octet.

would find this plot familiar; it resembles the *weight diagram* for the octet representation of SU(3). This suggests that the strong interactions which produce the observed particle supermultiplets have a symmetry appropriate to this group.[1]

We can also make a similar diagram for the spin 0^- mesons (Fig. 8-2). When the scheme for classifying the particles on the basis of SU(3) symmetry was proposed by Gell-Mann[2] and independently by Y. Ne'eman,[3] the η^0 had not yet been discovered. Its subsequent discovery was the first successful prediction of the scheme. The group SU(3) has other possible weight diagrams and corresponding multiplicities; some of these are shown in Fig. 8-3. Higher multiplicities such as a 27-tuplet can also occur.[4] However no particles (or resonances) whose quantum numbers cannot be accommodated in octets or decuplets have been found. (For example, a particle with $Y = -1$ and $T_3 = -1$ would not fit into a decuplet or octet.) Other supermultiplets will be discussed in Sec. 8-4.

[1] We emphasize that this symmetry is only approximate; otherwise the masses of all the particles in the supermultiplet would be equal (or equivalently, the energy levels would be degenerate).

[2] M. Gell-Mann, previously unpublished report included in M. Gell-Mann and Y. Ne'eman, "The Eightfold Way," Benjamin, New York, 1964. This volume also contains reprints of many of the articles cited in this chapter.

[3] Y. Ne'eman, *Nucl. Phys.* **26,** 222 (1961).

[4] In general, the allowed multiplicities for SU(3) multiplets are given by $N = \frac{1}{2}(j + 1)(j + k + 2)(k + 1)$, where $j = 0, 1, 2, \ldots$, and $k = 0, 1, 2, \ldots$.

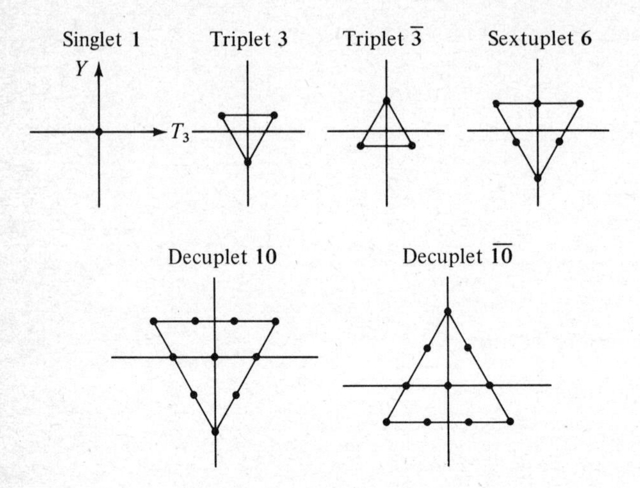

Figure 8-3 Other possible multiplets for the group SU(3).

8-3 SU(3) AND QUARKS

We now return to our model of SU(3), the building up of the supermultiplets from the three quarks. Historically the first attempt at such a model for the elementary particles was made by S. Sakata.[1] He identified the basic triplet with the p, n, and Λ. The Sakata model, however, did not agree with experiment (Prob. 8-3). The quark model was later introduced by Gell-Mann[2] and G. Zweig.[3] In analogy with the Sakata model, the quarks are designated the

[1] S. Sakata, *Progr. Theor. Phys.* (*Kyoto*) **16,** 686 (1956).

[2] M. Gell-Mann, *Phys. Letters* **8,** 214 (1964).

[3] G. Zweig, *CERN Reps.* 8182/TH 401 and 8419/TH 412 (1964), unpublished.

Table 8-1 Quantum numbers assumed for quarks

	p	$\bar{p}$	n	$\bar{n}$	λ	$\bar{\lambda}$
T, isospin	$\frac{1}{2}$	$\frac{1}{2}$	$\frac{1}{2}$	$\frac{1}{2}$	0	0
T_3	$\frac{1}{2}$	$-\frac{1}{2}$	$-\frac{1}{2}$	$\frac{1}{2}$	0	0
Y, hypercharge	$\frac{1}{3}$	$-\frac{1}{3}$	$\frac{1}{3}$	$-\frac{1}{3}$	$-\frac{2}{3}$	$\frac{2}{3}$
Q/e, charge	$\frac{2}{3}$	$-\frac{2}{3}$	$-\frac{1}{3}$	$\frac{1}{3}$	$-\frac{1}{3}$	$\frac{1}{3}$
S, strangeness	0	0	0	0	-1	$+1$
B, baryon number	$\frac{1}{3}$	$-\frac{1}{3}$	$\frac{1}{3}$	$-\frac{1}{3}$	$\frac{1}{3}$	$-\frac{1}{3}$

p, n, and λ, and their antiparticles the $\bar{p}$, $\bar{n}$, and $\bar{\lambda}$. The quantum numbers assumed for the quarks are shown in Table 8-1; the peculiar nature of these particles, in particular their fractional charge and baryon number, is apparent. This odd choice of quantum numbers is necessary to account for the observed supermultiplets. The quarks can be identified with the **3** and $\bar{\mathbf{3}}$ representations of SU(3), as can be seen from Fig. 8-4.

We now ask what states can be built up out of combinations of two or more quarks. We first take the simplest case, a quark-antiquark pair Q-$\bar{Q}$. A convenient graphic method for finding the resulting states is illustrated in Fig. 8-5. We start with a quark triplet and superimpose on each of the quarks an antiquark triplet. There are, of course, nine combinations, but three are superimposed at the origin. Except for the extra state at the origin, this looks like an SU(3) octet. In fact, it can be reduced to an octet plus a singlet. Symbolically,

$$\mathbf{3} \otimes \bar{\mathbf{3}} = \mathbf{8} \oplus 1$$

These multiplets correspond to meson multiplets since the associated baryon number is zero. The spin and parity quantum numbers are not specified by SU(3) and can have any value; however, it is assumed that all members of a multiplet have the same assignment.

If we combine two quarks by the same technique, we obtain a sextuplet and a triplet,

$$\mathbf{3} \otimes \mathbf{3} = \mathbf{6} \oplus \bar{\mathbf{3}}$$

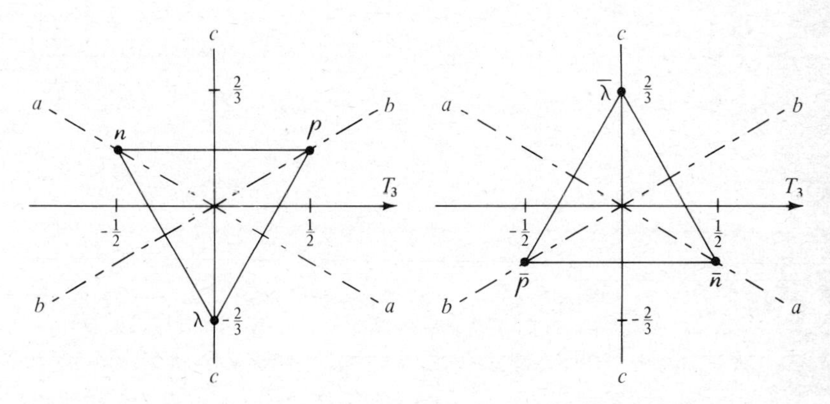

Figure 8-4 The quark and antiquark triplets.

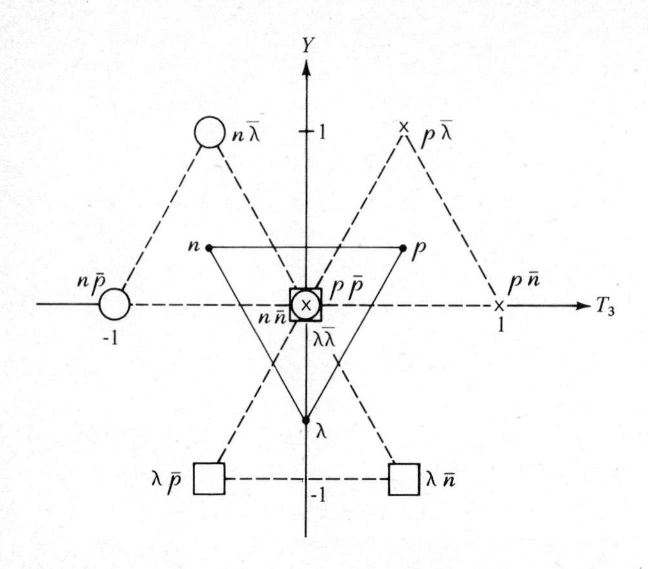

Figure 8-5 Graphic method for finding possible quark-antiquark combinations. An antiquark triplet is superimposed on each of the members of the quark triplet (solid circles).

These have baryon number $\frac{2}{3}$ and fractional charge, and do not appear to correspond to anything physical.

For the baryons, we need to combine three (or more) quarks to get a baryon number $+1$. We can start with the Q-Q sextuplet or triplet just mentioned and add another quark. From the sextuplet we get a decuplet and an octet (Prob. 8-1).

$$\mathbf{6} \otimes \mathbf{3} = \mathbf{10} \oplus \mathbf{8}$$

Starting with the triplet, we get an octet and singlet. Altogether

$$\mathbf{3} \otimes \mathbf{3} \otimes \mathbf{3} = \mathbf{10} \oplus \mathbf{8} \oplus \mathbf{8} \oplus \mathbf{1}$$

The simplest combinations (Q-$\bar{Q}$ for mesons and Q-Q-Q for baryons) therefore lead to meson octets or singlets and baryon decuplets, octets, or singlets. This seems to agree well with observation since no meson decuplets or baryon multiplets beyond the decuplet have been found. In the context of a quark model this would indicate that Q-Q-$\bar{Q}$-$\bar{Q}$ and Q-Q-Q-Q-$\bar{Q}$ states are unlikely or do not occur (see Prob. 8-4).

8-4 THE BARYON $\frac{3}{2}^+$ DECUPLET AND OTHER SUPERMULTIPLETS

When Gell-Mann and Ne'eman first proposed their scheme, the Ω^- had not yet been discovered. A $\frac{3}{2}^+$ baryon supermultiplet, containing the $\Delta(1236)$ and two resonances $\Sigma(1385)$ and $\Xi(1530)$ with strangeness of -1 and -2, seemed to exist, but its $Y = -2$ member was missing (Fig. 8-6). The missing particle would have strangeness of -3, and its mass could be predicted to be approximately 1,676 MeV, as discussed in the next section. Its mass would then be too small to permit its strong decay to $K + \Xi$, and it could therefore decay only through a weak interaction, most likely into a $\Xi + \pi$. Since the Ξ then decays into a Λ^0, which in turn decays into a nucleon, the resulting cascade of decays would give a dramatic signature in a bubble chamber. The prediction of the existence of this particle by Gell-Mann[1] started a race to find it. The race was won by a Brookhaven National Laboratory group[2]

[1] M. Gell-Mann, "Proceedings of the 1962 International Conference on High-energy Physics," p. 805, CERN, CERN Scientific Information Service, Geneva, 1962).

[2] V. Barnes et al., *Phys. Rev. Letters* **12,** 204 (1964). See also W. B. Fowler and N. P. Samios, *Sci. Am.* **211,** 36 (October, 1964).

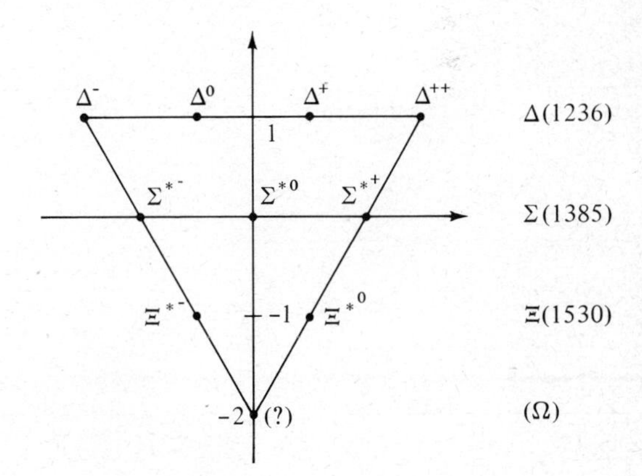

Figure 8-6 The $\frac{3}{2}^+$ baryon decuplet. The $Y = -2$ member was unknown when Gell-Mann and Ne'eman proposed that the hadrons were members of SU(3) multiplets.

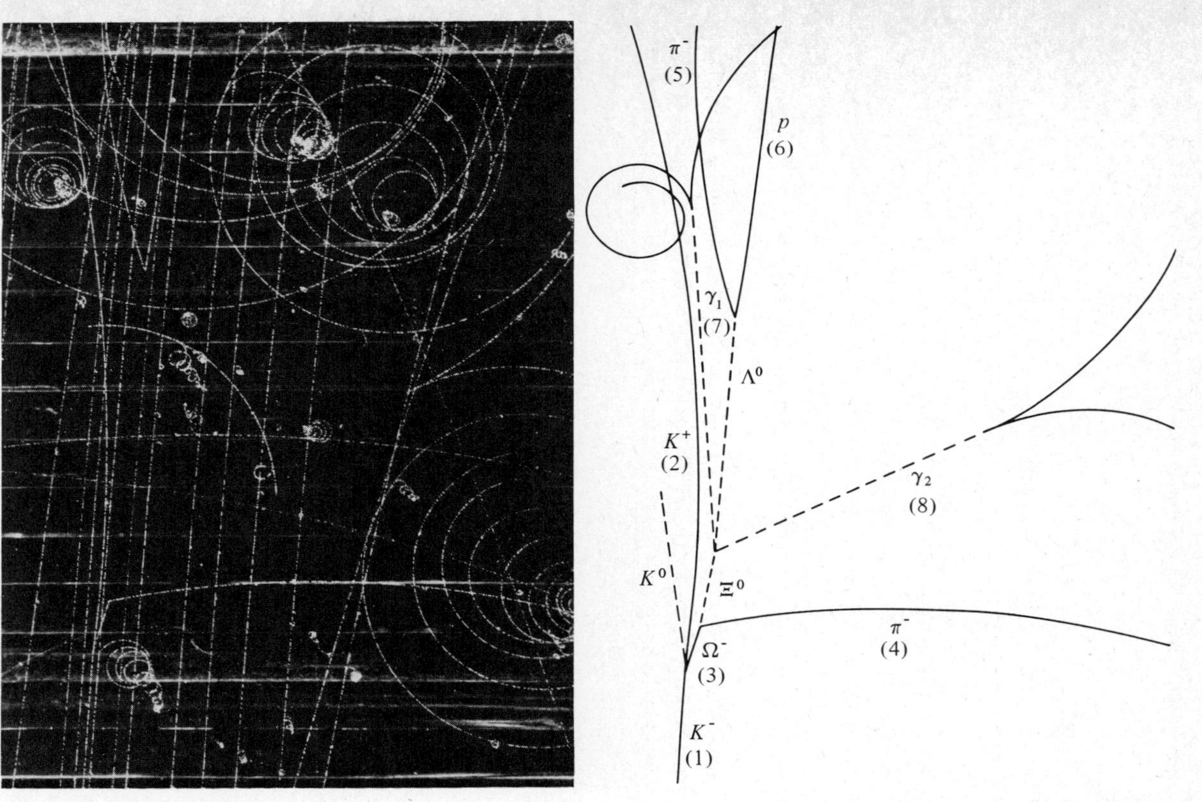

Figure 8-7 The first Ω^- event discovered by Barnes et al. in the Brookhaven National Laboratory 80-in. hydrogen bubble chamber. The legend on the right identifies the tracks. Track 1 is the incoming 5 GeV/c K^-. It produces the Ω^- through the reaction $K^- + p \rightarrow \Omega^- + K^0 + K^+$. The K^0 is not observed. The Ω^- decays into $\Xi^0 + \pi^-$, with the Ξ^0 in turn decaying into $\Lambda^0 + \pi^0$. The Λ^0 decays into $p + \pi^-$, and the π^0 into two γ's which both produce e^+-e^- pairs. (*From Brookhaven National Laboratory.*)

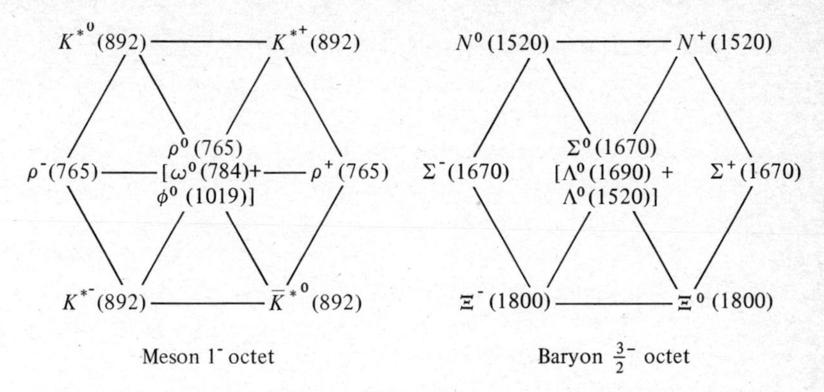

Figure 8-8 A meson 1^- octet and probable baryon $\frac{3}{2}^-$ octet. All members of both octets are short-lived states which decay by strong interactions (Table 3-1).

who found a bubble-chamber event interpreted as an Ω^-, which is shown in Fig. 8-7. The presently accepted mass of the Ω^- is about 1,672.5 MeV, which is in excellent agreement with the predicted mass of 1,676 MeV.

The prediction of the existence of the Ω^- and of its mass was the most dramatic success of SU(3). However, on the whole, the record of SU(3) is rather spotty. Unfortunately its predictive power without extra assumptions is limited, so that it is hard to test it directly. We shall discuss some of its predictions in the next section.

In addition to the supermultiplets given previously, several more can be completed with known resonances. A 1^- meson octet and a probable $\frac{3}{2}^-$ baryon octet are shown in Fig. 8-8. All states observed so far can be assigned to decuplets, octets, or singlets, with no evidence for 27-tuplets. However the spin-parity assignments of many of the resonances are uncertain, and many resonances whose assignments are known can only be fit into incomplete supermultiplets or singlets.

8-5 THE GELL-MANN–OKUBO MASS FORMULA AND OTHER SU(3) PREDICTIONS

In addition to classifying the particles, it is possible to do much more with SU(3), but, as noted earlier, this requires additional assumptions about the interaction which breaks the symmetry and causes the difference in masses between the supermultiplet members. The relatively small splitting in mass along the T_3 axis is presumably due to the charge-dependent electromagnetic interactions. The splitting along the Y direction is typically $\gtrsim 15$ percent of the mass, and must be due to some SU(3)-violating part of the strong interaction which depends on hypercharge. A statement of isospin invariance is that the strong interactions are invariant with respect to rotations in isospin space (which, for example, interchange proton and neutron). Such rotations are appropriate to the group SU(2). Similarly SU(3) invariance is equivalent to assuming that the strong interactions are invariant with respect to the interchange of any pair of the three quarks $\mathfrak{p}$, $\mathfrak{n}$, and λ.[1]

The effect of the SU(3)-violating interaction in breaking the degeneracy in mass of the supermultiplets is completely analogous to the Zeeman effect in atomic physics. If an atom is placed in a magnetic field B, a particular direction in coordinate space is singled out. This removes the degeneracy of the energy levels with respect to the direction of the total angular momentum J of the atom. The resultant splitting of the energy levels can be calculated from first-order perturbation theory as

$$\Delta E \propto \mathbf{M} \cdot \mathbf{B}$$

where $\mathbf{M}$ is the magnetic moment.

It is possible to obtain a mass formula for the SU(3) supermultiplets in an analogous fashion.[2] If the simplest possible form is assumed for the SU(3)-violating interaction and first-order perturbation theory is used,[3] the so-called Gell-Mann–Okubo mass formula is obtained

$$\bar{m} = m_0 + m_1 Y + m_2[T(T+1) - \tfrac{1}{4}Y^2] \qquad (8\text{-}1)$$

[1] This is equivalent to rotation about any one of the perpendicular bisectors aa, bb, or cc in Fig. 8-4. The rotation which interchanges $\mathfrak{p}$ and $\mathfrak{n}$ (or $\bar{\mathfrak{p}}$ and $\bar{\mathfrak{n}}$) corresponds to i-spin (isospin) rotation; that which interchanges $\mathfrak{n}$ and λ corresponds to a new symmetry, often referred to as *u spin*; and the interchange of $\mathfrak{p}$ and λ corresponds to another new symmetry *v spin*. Because of the assumed equivalence of the three quarks, all weight diagrams must be symmetrical about the axes aa, bb, and cc.

[2] S. Okubo, *Progr. Theor. Phys.* (Kyoto) **27**, 949 (1962).

[3] Despite the fact that for the meson octet the mass splitting is of the same order as the masses.

Here $\bar{m}$ is the average mass of the isospin multiplet, and m_0, m_1, and m_2 are constants within a given supermultiplet. If we use the particle symbol to represent the mass, Eq. (8-1) gives, for the baryon octet,

$$N = m_0 + m_1 + \tfrac{1}{2}m_2 \qquad \Lambda = m_0$$
$$\Xi = m_0 - m_1 + \tfrac{1}{2}m_2 \qquad \Sigma = m_0 + 2m_2$$

This leads to the relation

$$N + \Xi = \tfrac{1}{2}(3\Lambda + \Sigma) \tag{8-2}$$

Using the observed masses, we obtain 2,256 MeV for the left-hand side of the equation and 2,270 MeV for the right-hand side, in good agreement with the prediction.

If we apply Eq. (8-1) to the baryon $\frac{3}{2}^+$ decuplet, we obtain the *equal-spacing rule* (Prob. 8-2).

$$\begin{aligned} \Omega(1672) - \Xi(1530) &= \Xi(1530) - \Sigma(1385) \\ &= \Sigma(1385) - \Delta(1236) \end{aligned} \tag{8-3}$$

Again the agreement is relatively good. As mentioned earlier, the use of this formula to predict the mass of the Ω was the first great success of the theory.

If we now consider the meson 0^- octet, the equivalent of Eq. (8-2) is

$$2K = \tfrac{1}{2}(3\eta + \pi) \tag{8-4}$$

This formula was originally used to predict the η mass to be 615 MeV. It was actually found at 550 MeV, and the observation was made that the mass formula works much better if the mass *squared* is used instead of the mass for meson multiplets. The formula then predicts the η mass to be 567 MeV. The only proposed justification for the use of m^2 for bosons is that the wave equation for bosons involves m^2, while that for fermions involves m.

When we consider the meson 1^- octet (Fig. 8-8), the mass formula fails badly whether we use m or m^2 and whether we choose ω or ϕ as the central member. This led to the suggestion that because of the proximity in masses of the ω and ϕ, one of which is assumed to be an SU(3) singlet and the other a member of an octet with the same quantum numbers, the two states "mix" so that the physical particles are linear combinations of the SU(3) singlet and the central member of the octet. The nine members should therefore be considered a *nonet.*

With additional assumptions it is possible to make further predictions with SU(3). If the strong interactions had complete SU(3) symmetry, the degeneracy of the supermultiplets would be broken only by the electromagnetic interactions. We should then expect members of like charge to have the same electromagnetic properties. This leads directly to a prediction for the magnetic moments of members of the baryon octet

$$\begin{aligned} \mu(p) &= \mu(\Sigma^+) \\ \mu(\Sigma^-) &= \mu(\Xi^-) \end{aligned}$$

and

$$\mu(n) = \mu(\Xi^0)$$

Experimentally

$$\begin{aligned} \mu(p) &= 2.79 \\ \mu(n) &= -1.91 \\ \mu(\Sigma^+) &= 2.59 \pm 0.46 \end{aligned}$$

in units $e\hbar/2m_pc$, so that the first relation is satisfied within the large experimental certainty. By a similar argument, a relation between the electromagnetic mass differences of the charge multiplets can be developed which also agrees with that of the observed mass differences (within fairly large errors).[1] Some relations between cross sections have also been developed based on SU(3). In some cases, the predictions are relatively good; in others, they fail disastrously.

In addition to SU(3) symmetry there also exists a somewhat independent "quark model,"[2] in which the quark picture of elementary particles is taken rather literally. This model leads to relations between cross sections, magnetic moments, etc., some of which are independent of SU(3). For example, a very simple-minded comparison of meson-nucleon and nucleon-nucleon total cross sections in the quark model (Fig. 8-9) leads to the prediction that at very

[1] See, for example, D. B. Lichtenberg, "Unitary Symmetry and Elementary Particles," Academic, New York, 1970.

[2] See, for example, the articles by L. van Hove and F. Scheck in "Symmetry Principles and Fundamental Particles," Freeman, San Francisco, 1967.

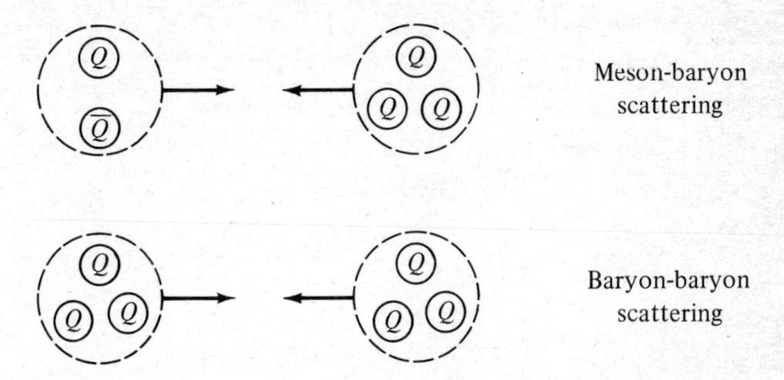

Figure 8-9 Meson-baryon and baryon-baryon scattering in the quark model. If all Q-Q and $\bar{Q}$-Q cross sections are approximately equal at high energies and shadowing is neglected, the meson-baryon total cross sections should be two-thirds those for baryon-baryon scattering.

high energies (when the Q-Q and $\bar{Q}$-Q total cross sections are expected to be asymptotically equal),

$$\begin{aligned}\sigma_T(\pi\text{-}p) &\approx \sigma_T(K\text{-}p)\\ &\approx \tfrac{2}{3}\sigma_T(N\text{-}N)\end{aligned}$$

This prediction neglects effects due to multiple scattering and "shadowing" of one quark by another. Experimentally at 55 GeV, the highest energy for which complete data are presently available,

$$\begin{aligned}\sigma_T(\pi\text{-}N) &\approx 23 \text{ mb}\\ \sigma_T(K\text{-}N) &\approx 18 \text{ mb}\end{aligned}$$

and

$$\sigma_T(N\text{-}N) \approx 38 \text{ mb}$$

in rough agreement with the prediction. In general, however, the agreement between the quark-model predictions and experiment is relatively poor.[1]

Even aside from these difficulties and the unanswered question of the existence of physical quarks (see next section), the quark model has significant problems. For example, it is not clear why we do not observe bound states of two, four, or five quarks or of two

[1] V. Barger and L. Durand, III, *Phys. Rev.* **156,** 1525 (1967).

quarks and an antiquark. Another problem concerns their mass. If their mass is relatively low ($\lesssim 1$ GeV), it is hard to understand why searches for low-mass quarks have been unsuccessful. On the other hand, if the hadrons are tightly bound composites of very massive quarks, it is hard to see why quark-model predictions based on the implicit assumption that the quarks are loosely bound (such as the prediction for total cross sections just mentioned) seem to work reasonably well.

8-6 DO QUARKS EXIST?

We have considered quarks the basis of a convenient model for SU(3). They may be no more than that, but on the other hand, there is the intriguing possibility that physical quarks exist. If so, they must be stable as a consequence of charge conservation.[1] Since they interact strongly, if they exist, it should be possible to produce them with significant cross sections, given enough energy. If the quark mass were $\lesssim 1$ GeV, their existence would probably have become apparent even before specific searches for them were made. We therefore suppose them to be relatively massive so that the baryons and mesons are very tightly bound structures with a binding energy comparable to the sum of the masses of the constituent quarks.

As might be expected, many experiments have been done to search for quarks. These can be divided roughly into two classes:

1. Searches for quarks trapped in matter
2. Searches for quarks in flight

In the first type of experiment, bulk matter is studied to find quarks which have come to rest and are bound in atoms or molecules. These might have been formed when the elements were created or as a result of bombardment by cosmic radiation. Most searches utilize some means of concentrating the quarks, and their interpre-

[1] They may however be tightly bound to another particle with unit charge to form a composite with charge $\frac{4}{3}$, or they may decay into a baryon plus antidiquark. In any case some particle with fractional charge must be absolutely stable.

tation requires assumptions about the chemistry of quarks. It is thought that negative quarks would be captured in Bohr orbits by nuclei. In this case, because of the great mass of the quarks, the radius of their orbits would be comparable to the nuclear radius. Since the chemical properties of an atom are determined by its outer electrons, an atom containing a quark would differ only slightly in its chemical behavior from an ordinary one. Positive quarks, on the other hand, would probably capture an electron and behave like an oversize hydrogen atom with excess negative charge.

As an example of an experiment to search for quarks trapped in matter, we consider briefly the experiment of Chupka, Schiffer, and Stevens.[1] They evaporated large quantities of seawater and passed the vapor between charged plates. Negative ions from the plates were further concentrated on a positively charged filament. The charges of the ions on the filament were then studied by observing the behavior of the evaporation rate when a potential was applied to the filament. Similar techniques were used for meteorite, dust, and air samples. The interpretation of the results is difficult. Using estimates of quark production by cosmic rays, the authors arrived at a lower limit on the quark mass of about 10 GeV; however this estimate might be quite optimistic.[2]

Searches for quarks in flight have been carried out with cosmic rays and with beams from high-energy accelerators. Most of these make use of the fact that the energy loss of fractionally charged particles in matter is considerably less than that of particles with unit charge. For high-energy particles $dE/dx \propto q^2$, so that on the average a quark with charge $e/3$ loses one-ninth as much energy as a particle with charge e, and a quark with charge $2e/3$ loses four-ninths as much. However the statistical fluctuations in the energy loss over short path lengths are large enough that many samplings must be made to obtain adequate separation. A fairly typical experiment of this sort is that of R. Gomez et al.,[3] who used an array

[1] W. Chupka, J. Schiffer, and C. Stevens, *Phys. Rev. Letters* **17,** 60 (1966). Aside from the interesting physics, this paper is worth reading for its entertainment value.

[2] Other types of quark searches in bulk matter are also discussed by Chupka, Schiffer, and Stevens. See also D. M. Rank, *Phys. Rev.* **176,** 1635 (1968).

[3] R. Gomez, H. Kobrak, A. Moline, J. Mullins, C. Orth, J. Van Putten, and G. Zweig, *Phys. Rev. Letters* **18,** 1022 (1967).

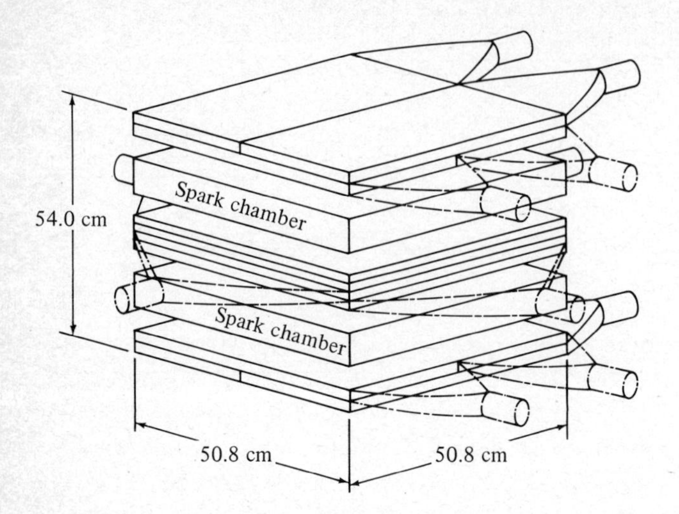

Figure 8-10 The apparatus used by Gomez et al. to search for fractionally charged particles in cosmic rays.

of scintillation counters and spark chambers to search for fractionally charged particles in cosmic rays. Their apparatus is shown in Fig. 8-10. The energy loss of particles traversing the apparatus was sampled by six pairs of scintillation counters. The spark chambers were triggered when particles with unusually low energy loss were detected and served to verify that a single particle traversed the sensitive volume. In the experiment approximately 1.5×10^8 cosmic-ray particles traversed the apparatus, and no quark candidates were found.

The only quark search which has yielded a suggestion of a positive result is that of C. McCusker and coworkers.[1] In a cloud-chamber experiment to study the cores of large showers produced in the upper atmosphere by very energetic cosmic rays, they found several examples of lightly-ionizing particles which could be quarks with charge $2e/3$. However there are many objections to the interpretation of C. McCusker and his coworkers,[2] and similar experiments with

[1] C. McCusker and I. Cairns, *Phys. Rev. Letters* **23,** 658 (1969); I. Cairns, C. McCusker, L. Peak, and R. Woolcott, *Phys. Rev.* **186,** 1394 (1969).

[2] R. Adair and H. Kasha, *Phys. Rev. Letters* **23,** 1355 (1969); D. Rahm and R. Louttit, *Phys. Rev. Letters* **24,** 278 (1970).

larger detectors and more sophisticated apparatus have thus far yielded only negative results.[1]

PROBLEMS

8-1 Show graphically that an SU(3) sextuplet when combined with a quark triplet gives a decuplet and an octet, i.e.,

$$\mathbf{6} \otimes \mathbf{3} = \mathbf{10} \oplus \mathbf{8}$$

8-2 Derive the equal-spacing rule, Eq. (8-3), from the Gell-Mann–Okubo mass formula. Derive Eq. (8-2) from the mass formula.

8-3 The original Sakata model used two baryons and an antibaryon to generate the baryon supermultiplets. Show that this does not generate an octet; i.e., show that $\mathbf{3} \otimes \mathbf{3} \otimes \bar{\mathbf{3}}$ does not contain an **8**.

8-4 Particles with a quark composition other than Q-Q-Q or Q-$\bar{Q}$ are considered "exotic." Specify a possible combination of quarks to give a particle with baryon number $+1$, strangeness $+1$, and charge $+e$ (the "Z^*"). Specify a combination of quarks for a meson with zero strangeness and charge of $+2e$. Consult the latest *Review of Particle Properties*[2] for evidence regarding the existence of such particles.

[1] A. F. Clark et al., *Phys. Rev. Letters* **27,** 51 (1971).

[2] These are published biannually, usually in *Physics Letters* or *Reviews of Modern Physics.* The latest to date appears in *Phys. Letters* **B39** (April, 1972).

CHAPTER 9

epilogue

High-energy physics today is an experimental science. We are exploring unknown modes of behavior of matter under completely novel conditions. The field has all the excitement of new discoveries in a virgin land, full of hidden treasures, the hoped-for fundamental insights into the structure of matter. It will take some time before we can produce a rational map of that new land.

Victor F. Weisskopf,
Comments on Nucl. Part. Phys. **3,** 1 (January, 1969)

Despite the tremendous progress in understanding the elementary particles and their interactions that has been made in the past several decades, many very basic questions remain to be answered. Why are there apparently four classes of interactions, not more or less, and how are these classes related to each other? Why is there a muon? How can the hadron mass spectrum be explained? Why are there no baryons with positive strangeness? Do quarks exist? Do magnetic monopoles exist, and if not, why not? What is the significance of the small CP violation? Why is the electric charge of the proton equal to that of the electron to better than one part in 10^{19}? The list goes on and on.

Are we even asking the right questions? Probably not. We are deluged with experimental data and can hardly see the forest for the trees. It is likely that men will some day look back at us and marvel at how naïve we were. Let us hope so!

Finally we ask, What is the significance of it all? Elementary particle research is the spearhead of all science: it explores the most fundamental questions man can ask and hope to answer by scientific inquiry. Where it will take us (and when we shall get there) cannot be foreseen. But when we look back and see the far-reaching significance of previous research along these lines, it is hard to believe that nature does not have many more secrets of incalculable importance waiting to be discovered.

PROBLEM

9-1 Answer the questions posed at the beginning of Chap. 9.

APPENDIX A

physical and numerical constants

PHYSICAL CONSTANTS

N_0 $= 6.022169\ (40) \times 10^{23}$ mole^{-1} (based on $A_{C^{12}} = 12$)
c $= 2.9979250\ (10) \times 10^{10}$ cm/sec
e $= 4.803250\ (21) \times 10^{-10}$ esu $= 1.6021917\ (70) \times 10^{-19}$ C
1 MeV $= 1.6021917\ (70) \times 10^{-6}$ erg
$\hbar$ $= 6.582183\ (22) \times 10^{-22}$ MeV-sec
$= 1.0545919\ (80) \times 10^{-27}$ erg-sec
$\hbar c$ $= 1.9732891\ (66) \times 10^{-11}$ MeV-cm
$= 197.32891\ (66)$ MeV-fm
$= 0.6240088\ (21)$ GeV-mb$^{\frac{1}{2}}$
α $= e^2/\hbar c = 1/137.03602\ (21)$
$k_{\text{Boltzmann}}$ $= 1.380622\ (59) \times 10^{-16}$ erg/°K
$= 8.61708\ (37) \times 10^{-11}$ MeV/°K $= 1$ eV/11604.85 (49)°K
m_e $= 0.5110041\ (16)$ MeV $= 9.109558\ (54) \times 10^{-31}$ kg
m_p $= 938.2592\ (52)$ MeV $= 1836.109\ (11)\ m_e$
$= 6.72211\ (63)\ m_{\pi^\pm}$
$= 1.00727661\ (8)\ m_1$ [where $m_1 = 1$ amu $= \frac{1}{12} m_{C^{12}} =$ 931.4812 (52) MeV]
r_e $= e^2/m_e c^2 = 2.817939\ (13)$ fm (1 fm $= 10^{-13}$ cm)
$\lambda\!\!\!^{-}_e$ $= \hbar/m_e c = r_e \alpha^{-1} = 3.861592\ (12) \times 10^{-11}$ cm
$a_{\infty\ \text{Bohr}}$ $= \hbar^2/m_e e^2 = r_e \alpha^{-2} = 0.52917715\ (81)$ Å (1Å $= 10^{-8}$ cm)
σ_{Thomson} $= \frac{8}{3}\pi r_e^2 = 0.6652453\ (61) \times 10^{-24}$ cm^2 $= 0.6652453\ (61)$ b
μ_{Bohr} $= e\hbar/2m_e c = 0.5788381\ (18) \times 10^{-14}$ MeV/G
μ_{nucleon} $= e\hbar/2m_p c = 3.152526\ (21) \times 10^{-18}$ MeV/G
$\frac{1}{2}\omega^e_{\text{cyclotron}}$ $= e/2m_e c = 8.794014\ (27) \times 10^6$ rad/sec-G
$\frac{1}{2}\omega^p_{\text{cyclotron}}$ $= e/2m_p c = 4.789484\ (27) \times 10^3$ rad/sec-G

Hydrogenlike atom (nonrelativistic, μ = reduced mass):

$$\left.\frac{v}{c}\right|_{\mathrm{rms}} = \frac{Ze^2}{n\hbar c} \qquad E_n = \frac{\mu}{2}v^2 = \frac{\mu Z^2 e^4}{2(n\hbar)^2} \qquad a_n = \frac{n^2\hbar^2}{\mu Ze^2}$$

$R_\infty = m_e e^4/2\hbar^2 = m_e c^2\alpha^2/2$ = 13.605826 (45) eV (Rydberg)
$pc = 0.3\ H\rho$ (MeV, kG, cm); 0.3 (which is $10^{-11}c$) enters because there are ≈300 V/esu V.

1 year (sidereal)	= 365.256 days = 3.1558×10^7 sec ($\approx \pi \times 10^7$ sec)
Density of dry air	= 1.205 mg/cm³ (at 20°C, 760 mm)
Acceleration by gravity	= 980.62 cm/sec² (sea level, 45°)
Gravitational constant	= 6.6732 (31) $\times 10^{-8}$ cm³/g-sec²
1 cal (thermochemical)	= 4.184 joules
1 atm	= 1,033.2275 g/cm²
1 eV per particle	= 11,604.85 (49) °K (from $E = kT$)

NUMERICAL CONSTANTS

π = 3.1415927	1 rad = 57.2957795°	$\sqrt{\pi}$ = 1.7724539
e = 2.7182818	$1/e$ = 0.3678794	$\sqrt{2}$ = 1.4142136
ln 2 = 0.6931472	ln 10 = 2.3025851	$\sqrt{3}$ = 1.7320508
log 2 = 0.3010300	log e = 0.4342945	$\sqrt{10}$ = 3.1622777

SOURCE: The Particle Data Group, The Review of Particle Properties, *Rev Mod. Phys. Suppl.*, April, 1971. [After the compilation by Stanley J. Brodsky, based mainly on the adjustment of the fundamental physical constants by B. N. Taylor, W. H. Parker, and D. N. Langenberg, *Rev. Mod. Phys.* **41**, 375 (1969).] The figures in parentheses correspond to the 1 standard deviation uncertainty in the last digits of the main number.

APPENDIX B

summary of relativistic kinematics

Units are chosen so that $c \equiv 1$.

SINGLE PARTICLES[1]

$$\gamma \equiv (1 - \beta^2)^{-\frac{1}{2}} \tag{B-1}$$

Momentum: $p = m\beta\gamma$

Total energy: $E = \sqrt{p^2 + m^2} = m\gamma$

Kinetic energy: $T = E - m$

It is convenient to define relativistic quantities in terms of four-vectors, which transform under the Lorentz transformation much as ordinary vectors in three-dimensional space. We shall designate four-vectors with a bar over the symbol (for example, $\bar{p}$) and ordinary vectors in boldface.

Thus

$$\mathbf{r} = \begin{pmatrix} x \\ y \\ z \end{pmatrix} \qquad \mathbf{p} = \begin{pmatrix} p_x \\ p_y \\ p_z \end{pmatrix}$$

$$\bar{r} = \begin{pmatrix} x \\ y \\ z \\ it \end{pmatrix} \equiv (\mathbf{r},t) \qquad \bar{p} = \begin{pmatrix} p_x \\ p_y \\ p_z \\ iE \end{pmatrix} \equiv (\mathbf{p},t) \tag{B-2}$$

[1] For a more detailed discussion of the topics of this appendix, see R. Hagedorn, "Relativistic Kinematics," Benjamin, New York, 1963.

Four-vectors have the following important properties:

1. The scalar product of 2 four-vectors $\bar{a}$ and $\bar{b}$ is

$$\bar{a} \cdot \bar{b} = a_x b_x + a_y b_y + a_z b_z - a_t b_t \tag{B-3}$$

2. The scalar product of any 2 four-vectors is an invariant, for example,

$$\begin{aligned} \bar{p} \cdot \bar{p} &= p_x^2 + p_y^2 + p_z^2 - E^2 \\ &= \mathbf{p} \cdot \mathbf{p} - E^2 \\ &= -m^2 \end{aligned} \tag{B-4}$$

   where m is the rest mass.
3. The transformation of a four-vector from one system to another that is moving relative to it is the Lorentz transformation,

$$\begin{pmatrix} a_x \\ a_y \\ a_z \\ ia_t \end{pmatrix} = \begin{pmatrix} 1 & 0 & 0 & 0 \\ 0 & 1 & 0 & 0 \\ 0 & 0 & \gamma & \gamma\beta \\ 0 & 0 & \gamma\beta & \gamma \end{pmatrix} \begin{pmatrix} a_x^* \\ a_y^* \\ a_z^* \\ ia_t^* \end{pmatrix} \tag{B-5}$$

   where the starred system is moving with velocity $+\beta$ along the Z axis, and $\gamma = (1 - \beta^2)^{-\frac{1}{2}}$.

MANY-PARTICLE SYSTEMS

A group of several particles has a total four-momentum vector

$$\bar{p} = \bar{p}_1 + \bar{p}_2 + \cdots$$

That is,

$$\begin{aligned} \mathbf{p} &= \mathbf{p}_1 + \mathbf{p}_2 + \cdots \\ E &= E_1 + E_2 + \cdots \end{aligned}$$

The *center-of-mass system* (c.m.s.) is the frame in which the group of particles has zero total momentum; that is, $\mathbf{p} = 0$. The rest mass of the system (also called the total c.m.s. energy or effective mass), μ, is given by [Eq. (B-4)].

$$\begin{aligned}\mu^2 &= -\bar{p} \cdot \bar{p} \\ &= E^2 - p^2 \\ &= \left(\sum E_i\right)^2 - \left|\sum \mathbf{p}_i\right|^2\end{aligned}$$

The square of the total c.m.s. energy is often denoted by s. Because it is defined as the scalar product of 2 four-vectors, it is a relativistic invariant.

TWO-BODY COLLISIONS

If two particles of four-momentum $\bar{p}_1$ and $\bar{p}_2$ collide, the square of the total c.m.s. energy s is given by

$$\begin{aligned}s &= -(\bar{p}_1 + \bar{p}_2) \cdot (\bar{p}_1 + \bar{p}_2) \\ &= -\bar{p}_1{}^2 - \bar{p}_2{}^2 - 2\bar{p}_1 \cdot \bar{p}_2 \\ &= m_1{}^2 + m_2{}^2 + 2E_1E_2 - 2\mathbf{p}_1 \cdot \mathbf{p}_2\end{aligned} \tag{B-6}$$

If particle 2 is at rest in the laboratory,

$$\begin{aligned}\sqrt{s} &= \sqrt{m_1{}^2 + m_2{}^2 + 2E_1M_2} \\ &= \sqrt{(m_1 + m_2)^2 + 2T_1m_2}\end{aligned} \tag{B-6a}$$

where E_1 is the total energy of particle 1 in the laboratory, and T_1 its kinetic energy in the laboratory.

The velocity and the Lorentz factor of the c.m.s. in the laboratory are

$$\begin{aligned}\beta_c &= \frac{p_1}{E_1 + m_2} \\ \gamma_c &= \frac{E_1 + m_2}{s^{\frac{1}{2}}}\end{aligned} \tag{B-7}$$

The threshold for the reaction

$$1 + 2 \to 3 + 4 + 5 + \cdots$$

occurs when the total energy in the c.m.s. is equal to the total rest mass in the final state, or

$$\sqrt{s_0} = m_3 + m_4 + m_5 + \cdots$$

If we use Eq. (B-6*a*) for the target at rest in the laboratory, we find that the laboratory kinetic energy for particle 1 at threshold is

$$T_0 = -\frac{(m_1 + m_2 + m_3 + m_4 + \cdots)Q}{2m_2} \tag{B-8}$$

where

$$Q \equiv (m_1 + m_2) - (m_3 + m_4 + m_5 + \cdots)$$

Table B-1 gives the total energy in the c.m.s. and momentum of either particle in the c.m.s. for beams of various particles incident on protons (at rest in the laboratory).

Table B-1 Center-of-mass energy and momentum versus beam momentum

P Beam, MeV/c	C.M. Energy, MeV				Momentum in c.m., MeV/c			
	γp ep	πp	Kp	pp	γp ep	πp	Kp	pp
0	939	1078	1432	1877	0	0	0	0
20	958	1079	1432	1877	20	17	13	10
40	977	1083	1433	1877	38	35	26	20
60	996	1089	1434	1877	56	52	39	30
80	1015	1096	1436	1878	74	68	52	40
		T(PI) = PBEAM - 59 MEV						
100	1033	1105	1439	1879	91	85	65	50
120	1051	1116	1441	1880	107	101	78	60
140	1069	1127	1445	1882	123	117	91	70
160	1087	1139	1449	1883	138	132	104	80
180	1104	1152	1453	1885	153	147	116	90
		T(PI) = PBEAM - 92 MEV						
200	1121	1165	1457	1887	167	161	129	99
220	1137	1178	1462	1889	182	175	141	109
240	1154	1192	1468	1892	195	189	153	119
260	1170	1206	1474	1894	209	202	166	129
280	1186	1219	1480	1897	222	215	178	138
		T(PI) = PBEAM - 107 MEV						
300	1201	1233	1486	1900	234	228	189	148
320	1217	1247	1493	1903	247	241	201	158
340	1232	1261	1500	1906	259	253	213	167
360	1247	1274	1507	1910	271	265	224	177
380	1262	1288	1514	1913	282	277	235	186
		T(PI) = PBEAM - 115 MEV						
400	1277	1302	1522	1917	294	288	247	196
420	1292	1315	1530	1921	305	300	258	205
440	1306	1329	1538	1925	316	311	268	214
460	1320	1342	1546	1929	327	322	279	224
480	1335	1356	1554	1933	337	332	290	233
		T(PI) = PBEAM - 120 MEV						
500	1349	1369	1563	1938	348	343	300	242
520	1362	1382	1572	1943	358	353	310	251
540	1376	1395	1580	1947	368	363	321	260
560	1390	1408	1589	1952	378	373	331	269
580	1403	1421	1598	1957	388	383	341	278
		T(PI) = PBEAM - 123 MEV						
600	1416	1434	1607	1962	397	393	350	287
620	1430	1446	1616	1967	407	402	360	296
640	1443	1459	1625	1973	416	412	370	304
660	1456	1472	1634	1978	425	421	379	313
680	1468	1484	1644	1984	434	430	388	322
		T(PI) = PBEAM - 125 MEV						
700	1481	1496	1653	1989	443	439	397	330
720	1494	1509	1662	1995	452	448	406	339
740	1506	1521	1671	2001	461	457	415	347
760	1519	1533	1681	2007	470	465	424	355
780	1531	1545	1690	2013	478	474	433	364
		T(PI) = PBEAM - 127 MEV						
800	1543	1557	1699	2019	486	482	442	372
820	1555	1569	1709	2025	495	490	450	380
840	1567	1580	1718	2031	503	499	459	388
860	1579	1592	1728	2037	511	507	467	396
880	1591	1604	1737	2043	519	515	475	404
		T(PI) = PBEAM - 129 MEV						
900	1603	1615	1747	2049	527	523	483	412
920	1615	1627	1756	2056	535	531	492	420
940	1626	1638	1765	2062	542	538	500	428
960	1638	1649	1775	2069	550	546	508	435
980	1649	1661	1784	2075	558	554	515	443
		T(PI) = PBEAM - 130 MEV						
1000	1660	1672	1794	2082	565	561	523	451
1020	1672	1683	1803	2088	573	569	531	458
1040	1683	1694	1812	2095	580	576	538	466
1060	1694	1705	1822	2102	587	583	546	473
1080	1705	1716	1831	2108	594	591	553	481
		T(PI) = PBEAM - 131 MEV						
1100	1716	1726	1840	2115	601	598	561	488
1120	1727	1737	1850	2122	609	605	568	495
1140	1738	1748	1859	2129	616	612	575	502
1160	1748	1758	1868	2135	622	619	583	510
1180	1759	1769	1877	2142	629	626	590	517
		T(PI) = PBEAM - 131 MEV						
1200	1770	1780	1887	2149	636	633	597	524
1220	1780	1790	1896	2156	643	639	604	531
1240	1791	1800	1905	2163	650	646	611	538
1260	1801	1811	1914	2170	656	653	618	545
1280	1812	1821	1923	2177	663	660	624	552
		T(PI) = PBEAM - 132 MEV						
1300	1822	1831	1932	2184	669	666	631	559
1320	1832	1841	1941	2191	676	673	638	565
1340	1843	1851	1950	2198	682	679	645	572
1360	1853	1862	1959	2205	689	685	651	579
1380	1863	1872	1968	2212	695	692	658	585
		T(PI) = PBEAM - 133 MEV						
1400	1873	1881	1977	2219	701	698	664	592
1420	1883	1891	1986	2226	708	704	671	599
1440	1893	1901	1995	2233	714	711	677	605
1460	1903	1911	2004	2240	720	717	684	612
1480	1912	1921	2013	2247	726	723	690	618
		T(PI) = PBEAM - 133 MEV						
1500	1922	1930	2022	2254	732	729	696	624
1520	1932	1940	2031	2261	738	735	702	631
1540	1942	1950	2039	2268	744	741	709	637
1560	1951	1959	2048	2275	750	747	715	643
1580	1961	1969	2057	2282	756	753	721	650
		T(PI) = PBEAM - 133 MEV						
1600	1970	1978	2065	2289	762	759	727	656
1620	1980	1988	2074	2296	768	765	733	662
1640	1989	1997	2083	2304	773	770	739	668
1660	1999	2006	2091	2311	779	776	745	674
1680	2008	2016	2100	2318	785	782	751	680
		T(PI) = PBEAM - 134 MEV						
1700	2018	2025	2109	2325	791	788	756	686
1720	2027	2034	2117	2332	796	793	762	692
1740	2036	2043	2126	2339	802	799	768	698
1760	2045	2053	2134	2346	807	805	774	704
1780	2054	2062	2143	2353	813	810	779	710
		T(PI) = PBEAM - 134 MEV						
1800	2064	2071	2151	2360	818	816	785	716
1820	2073	2080	2159	2367	824	821	791	721
1840	2082	2089	2168	2374	829	827	796	727
1860	2091	2098	2176	2381	835	832	802	733
1880	2100	2107	2184	2388	840	837	808	739
		T(PI) = PBEAM - 134 MEV						
1900	2108	2115	2193	2395	845	843	813	744
1920	2117	2124	2201	2402	851	848	818	750
1940	2126	2133	2209	2409	856	853	824	756
1960	2135	2142	2217	2416	861	859	829	761
1980	2144	2150	2226	2423	867	864	835	767
		T(PI) = PBEAM - 135 MEV						
2000	2153	2159	2234	2430	872	869	840	772
2020	2161	2168	2242	2437	877	874	845	778
2040	2170	2176	2250	2444	882	879	851	783
2060	2179	2185	2258	2451	887	885	856	789
2080	2187	2194	2266	2458	892	890	861	794
		T(PI) = PBEAM - 135 MEV						
2100	2196	2202	2274	2465	897	895	866	799
2120	2204	2211	2282	2472	902	900	872	805
2140	2213	2219	2290	2479	907	905	877	810
2160	2221	2227	2298	2486	912	910	882	815
2180	2230	2236	2306	2493	917	915	887	821
		T(PI) = PBEAM - 135 MEV						
2200	2238	2244	2314	2500	922	920	892	826
2220	2246	2253	2322	2507	927	925	897	831
2240	2255	2261	2330	2514	932	930	902	836
2260	2263	2269	2338	2520	937	934	907	841
2280	2271	2277	2346	2527	942	939	912	846
		T(PI) = PBEAM - 135 MEV						
2300	2280	2286	2353	2534	947	944	917	852
2320	2288	2294	2361	2541	951	949	922	857
2340	2296	2302	2369	2548	956	954	927	862
2360	2304	2310	2377	2555	961	959	932	867
2380	2312	2318	2384	2561	966	963	937	872
		T(PI) = PBEAM - 135 MEV						
2400	2320	2326	2392	2568	970	968	941	877
2420	2328	2334	2400	2575	975	973	946	882
2440	2336	2342	2407	2582	980	977	951	887
2460	2344	2350	2415	2589	984	982	956	892
2480	2352	2358	2423	2595	989	987	960	897
		T(PI) = PBEAM - 136 MEV						
2500	2360	2366	2430	2602	994	991	965	901
2520	2368	2374	2438	2609	998	996	970	906
2540	2376	2382	2445	2616	1003	1001	975	911
2560	2384	2390	2453	2622	1007	1005	979	916
2580	2392	2398	2460	2629	1012	1010	984	921
		T(PI) = PBEAM - 136 MEV						

Table B-1 Center-of-mass energy and momentum versus beam momentum (*Continued*)

P Beam, MeV/c	C.M. Energy, MeV				Momentum in c.m., MeV/c			
	γp ep	πp	Kp	pp	γp ep	πp	Kp	pp
2600	2400	2405	2468	2636	1017	1014	988	926
2620	2408	2413	2475	2643	1021	1019	993	930
2640	2415	2421	2483	2649	1025	1023	998	935
2660	2423	2429	2490	2656	1030	1028	1002	940
2680	2431	2436	2498	2663	1034	1032	1007	944
		T(PI) = PBEAM – 136 MEV						
2700	2439	2444	2505	2669	1039	1037	1011	949
2720	2446	2452	2512	2676	1043	1041	1016	954
2740	2454	2459	2520	2682	1048	1045	1020	958
2760	2462	2467	2527	2689	1052	1050	1025	963
2780	2469	2474	2534	2696	1056	1054	1029	968
		T(PI) = PBEAM – 136 MEV						
2800	2477	2482	2542	2702	1061	1058	1034	972
2820	2484	2490	2549	2709	1065	1063	1038	977
2840	2492	2497	2556	2715	1069	1067	1042	981
2860	2499	2505	2563	2722	1074	1071	1047	986
2880	2507	2512	2570	2728	1078	1076	1051	990
		T(PI) = PBEAM – 136 MEV						
2900	2514	2520	2578	2735	1082	1080	1056	995
2920	2522	2527	2585	2742	1086	1084	1060	999
2940	2529	2534	2592	2748	1091	1088	1064	1004
2960	2537	2542	2599	2755	1095	1093	1069	1008
2980	2544	2549	2606	2761	1099	1097	1073	1013
		T(PI) = PBEAM – 136 MEV						

P Beam, GeV/c	C.M. Energy, GeV			Momentum in c.m., GeV/c		
	γp ep πp	Kp	pp	γp ep πp	Kp	pp
3.0	2.56	2.61	2.77	1.10	1.08	1.02
3.2	2.63	2.68	2.83	1.14	1.12	1.06
3.4	2.70	2.75	2.89	1.18	1.16	1.10
3.6	2.77	2.82	2.96	1.22	1.20	1.14
3.8	2.83	2.88	3.02	1.26	1.24	1.18
		T(PI) = PBEAM – .137 GEV				
4.0	2.90	2.95	3.08	1.29	1.27	1.22
4.2	2.96	3.01	3.14	1.33	1.31	1.26
4.4	3.03	3.07	3.19	1.36	1.34	1.29
4.6	3.09	3.13	3.25	1.40	1.38	1.33
4.8	3.15	3.19	3.31	1.43	1.41	1.36
		T(PI) = PBEAM – .138 GEV				
5.0	3.21	3.25	3.36	1.46	1.44	1.40
5.2	3.27	3.31	3.42	1.49	1.48	1.43
5.4	3.32	3.36	3.47	1.53	1.51	1.46
5.6	3.38	3.42	3.52	1.56	1.54	1.49
5.8	3.43	3.47	3.58	1.59	1.57	1.52
		T(PI) = PBEAM – .138 GEV				
6.0	3.49	3.52	3.63	1.61	1.60	1.55
6.2	3.54	3.58	3.68	1.64	1.63	1.58
6.4	3.59	3.63	3.73	1.67	1.65	1.61
6.6	3.65	3.68	3.78	1.70	1.68	1.64
6.8	3.70	3.73	3.83	1.73	1.71	1.67
		T(PI) = PBEAM – .138 GEV				
7.0	3.75	3.78	3.87	1.75	1.74	1.70
7.2	3.80	3.83	3.92	1.78	1.76	1.72
7.4	3.85	3.88	3.97	1.81	1.79	1.75
7.6	3.89	3.93	4.02	1.83	1.82	1.78
7.8	3.94	3.97	4.06	1.86	1.84	1.80
		T(PI) = PBEAM – .138 GEV				
8.0	3.99	4.02	4.11	1.88	1.87	1.83
8.2	4.04	4.07	4.15	1.91	1.89	1.85
8.4	4.08	4.11	4.20	1.93	1.92	1.88
8.6	4.13	4.16	4.24	1.95	1.94	1.90
8.8	4.17	4.20	4.29	1.98	1.96	1.93
		T(PI) = PBEAM – .138 GEV				
9.0	4.22	4.25	4.33	2.00	1.99	1.95
9.2	4.26	4.29	4.37	2.03	2.01	1.97
9.4	4.31	4.33	4.41	2.05	2.03	2.00
9.6	4.35	4.38	4.46	2.07	2.06	2.02
9.8	4.39	4.42	4.50	2.09	2.08	2.04
		T(PI) = PBEAM – .139 GEV				
10.0	4.43	4.46	4.54	2.12	2.10	2.07
10.5	4.54	4.57	4.64	2.17	2.16	2.12
11.0	4.64	4.67	4.74	2.22	2.21	2.18
11.5	4.74	4.77	4.84	2.28	2.26	2.23
12.0	4.84	4.86	4.93	2.33	2.31	2.28
		T(PI) = PBEAM – .139 GEV				
12.5	4.94	4.96	5.03	2.38	2.36	2.33
13.0	5.03	5.05	5.12	2.43	2.41	2.38
13.5	5.12	5.15	5.21	2.47	2.46	2.43
14.0	5.21	5.24	5.30	2.52	2.51	2.48
14.5	5.30	5.32	5.39	2.57	2.56	2.53
		T(PI) = PBEAM – .139 GEV				
15.0	5.39	5.41	5.47	2.61	2.60	2.57
16.0	5.56	5.58	5.64	2.70	2.69	2.66
17.0	5.73	5.75	5.81	2.78	2.77	2.75
18.0	5.89	5.91	5.97	2.87	2.86	2.83
19.0	6.05	6.07	6.12	2.95	2.94	2.91
		T(PI) = PBEAM – .139 GEV				
20.0	6.20	6.22	6.27	3.03	3.02	2.99
22.0	6.49	6.51	6.56	3.18	3.17	3.14
24.0	6.78	6.79	6.84	3.32	3.31	3.29
26.0	7.05	7.07	7.11	3.46	3.45	3.43
28.0	7.31	7.33	7.37	3.59	3.59	3.56
		T(PI) = PBEAM – .139 GEV				
30.0	7.56	7.58	7.62	3.72	3.71	3.69
32.0	7.81	7.82	7.86	3.85	3.84	3.82
34.0	8.04	8.06	8.10	3.97	3.96	3.94
36.0	8.27	8.29	8.33	4.08	4.08	4.06
38.0	8.50	8.51	8.55	4.20	4.19	4.17
		T(PI) = PBEAM – .139 GEV				
40.0	8.72	8.73	8.77	4.31	4.30	4.28
42.0	8.93	8.94	8.98	4.41	4.41	4.39
44.0	9.14	9.15	9.18	4.52	4.51	4.50
46.0	9.34	9.35	9.39	4.62	4.62	4.60
48.0	9.54	9.55	9.58	4.72	4.72	4.70
		T(PI) = PBEAM – .139 GEV				
50.0	9.73	9.74	9.78	4.82	4.81	4.80
55.0	10.20	10.21	10.25	5.06	5.05	5.04
60.0	10.65	10.66	10.69	5.28	5.28	5.26
65.0	11.08	11.10	11.12	5.50	5.50	5.49
70.0	11.50	11.51	11.54	5.71	5.71	5.69
		T(PI) = PBEAM – .139 GEV				
80.0	12.29	12.30	12.32	6.11	6.10	6.09
90.0	13.03	13.04	13.06	6.48	6.48	6.46
100.0	13.73	13.74	13.76	6.83	6.83	6.82
200.0	19.40	19.40	19.42	9.67	9.67	9.66
500.0	30.65	30.65	30.66	15.31	15.31	15.30
		T(PI) = PBEAM – .140 GEV				

SOURCE: Reproduced from the Particle Data Group, The Review of Particle Properties, *Rev. Mod. Phys. Suppl.* (April, 1971).

APPENDIX C

summary of terminology of group theory

A *group* is a set of abstract elements $a, b, c, \ldots$, finite or infinite in number, which obey the following rules:

1. A law of combination, usually called *multiplication,* exists such that every product of two elements and the square of every element is a member of the set.
2. The associative law holds, $a(bc) = (ab)c$.
3. The set contains a unit element e such that $ea = ae = a$ if a is any member of the set.
4. Every element a has an inverse a^{-1} so that $aa^{-1} = a^{-1}a = e$.

The set of all integers forms a group if the law of combination is addition (but not if it is multiplication). The unit element is then zero, and the inverse of an element k is $-k$.

Two groups G and H are said to be simply *isomorphic* if to each element $a, b, c, \ldots$ of G there corresponds an element $a', b', c', \ldots$ of H. If to every member of a group $a_1, a_2, \ldots$ we can associate a square matrix $A(a_1), A(a_2), \ldots$ in such a way that $a_i a_j = a_k$ implies $A(a_i)A(a_j) = A(a_k)$, then the matrices form a group isomorphous with the group $a_1, a_2, \ldots$. Such matrices are said to form a (*matrix*) *representation* of the group, with the order of the matrices termed the *dimension* of the representation. A representation is *reducible* if a new coordinate system can be found such that each of the matrices A can be written in the form

$$A(a_i) = \begin{pmatrix} A_1(a_i) & 0 \\ 0 & A_2(a_i) \end{pmatrix} \tag{C-1}$$

where A_1 is a matrix of lower order than the matrix A. An *irreducible* representation is one which cannot be reduced further.

Many of the groups important in physics are specified by one or more continuously varying parameters, for example, an angle of rotation θ or a translation a. Most of these have the property that if we form the product of any two group elements to get a third,

$$A(a_1)A(a_2) = A(a_3)$$

where a_1, a_2, and a_3 are particular values of the parameter, then a_3 is an analytic function of a_1 and a_2. A group specified by such an analytic function is called a *Lie group*. Examples of Lie groups are the sets of transformations associated with rotations or with Lorentz transformations.

Another example of a Lie group is the *general linear group* of dimension n, or GL(n), which can be written in the form

$$A = \begin{pmatrix} a_{11} & \cdots & a_{1n} \\ \cdots & \cdots & \cdots \\ a_{n1} & \cdots & a_{nn} \end{pmatrix} \tag{C-2}$$

where the parameters a_{ij} are, in general, complex. The group GL(n) is therefore in general characterized by $2n^2$ parameters. If we specialize to matrices whose determinant is unity (unimodular), we have the so-called *special* linear group SL(n). If we further consider only unitary matrices,[1] we have the special *unitary* group SU(n). The requirement of unitarity puts n^2 conditions on the $2n^2$ parameters, and with the requirement that the determinant be unity, we are left with $n^2 - 1$ parameters needed to characterize a member of the group SU(n).

Sophus Lie showed that most of the properties of continuous analytic groups (*Lie groups*) can be obtained by studying those elements which differ infinitesimally from the identity. In general, the properties of elements which can be reached continuously from

[1] The condition that a matrix be unitary is $AA\dagger = A\dagger A = E$, where E is the unit matrix and

$$A\dagger \equiv \begin{pmatrix} a_{11}^* & \cdots & a_{n1}^* \\ \cdots & \cdots & \cdots \\ a_{1n}^* & \cdots & a_{nn}^* \end{pmatrix}$$

Most matrices in quantum mechanics are unitary because of the requirement that probability be conserved. Rotations and translations in real space are also unitary because of the requirement that lengths of vectors be preserved.

the identity can be determined from the properties of matrix operators called the *generators* of the group. There are as many generators as there are parameters required to specify a group [that is, $n^2 - 1$ for SU(n)]. One possible choice for the generators of SU(2) is the so-called *Pauli spin matrices*

$$\begin{aligned}\sigma_1 &= \begin{pmatrix} 0 & 1 \\ 1 & 0 \end{pmatrix} \\ \sigma_2 &= \begin{pmatrix} 0 & -i \\ i & 0 \end{pmatrix} \\ \sigma_3 &= \begin{pmatrix} 1 & 0 \\ 0 & -1 \end{pmatrix}\end{aligned} \tag{C-3}$$

These matrices satisfy the commutation relations

$$(\sigma_\mu,\sigma_\nu) = 2i\epsilon_{\mu\nu\lambda}\sigma_\lambda \tag{C-4}$$

where $(\sigma_\mu,\sigma_\nu) \equiv \sigma_\mu\sigma_\nu - \sigma_\nu\sigma_\mu$ and $\epsilon_{\mu\nu\lambda} = \pm 1$ if the indices are an even (odd) permutation of 1, 2, 3. These commutation relations define the algebra of the group. The Pauli matrices are said to be a representation of the algebra for SU(2). These 2×2 matrices are in fact the lowest-dimensional representation of SU(2), but the group may also be represented by matrices of higher dimension.[1] The lowest-dimensional representation is called the *fundamental* representation and for SU(n) is n dimensional.

The fundamental representation of SU(3) therefore takes the form of eight 3×3 matrices analogous to the Pauli matrices.[2] Two of these can be written in diagonal form,

$$\begin{aligned}H_1 &= \frac{1}{\sqrt{6}}\begin{pmatrix} 1 & 0 & 0 \\ 0 & -1 & 0 \\ 0 & 0 & 0 \end{pmatrix} \\ H_2 &= \frac{1}{3\sqrt{2}}\begin{pmatrix} 1 & 0 & 0 \\ 0 & 1 & 0 \\ 0 & 0 & -2 \end{pmatrix}\end{aligned}$$

[1] In this case there are representations of dimension 2, 3, 4, ...

[2] Higher-dimensionality representations of SU(3) of dimension 6, 8, 10, . . . are possible.

These matrices act upon a set of three-dimensional basis vectors[1]

$$u_1 = \begin{pmatrix} 1 \\ 0 \\ 0 \end{pmatrix}$$

$$u_2 = \begin{pmatrix} 0 \\ 1 \\ 0 \end{pmatrix}$$

$$u_3 = \begin{pmatrix} 0 \\ 0 \\ 1 \end{pmatrix}$$

If we operate on these vectors with H_1 and H_2, we obtain the eigenvalue equations

$$H_1u_1 = \frac{1}{\sqrt{6}}\,u_1 \qquad H_2u_1 = \frac{1}{3\sqrt{2}}\,u_1$$

$$H_1u_2 = \frac{1}{-\sqrt{6}}\,u_2 \qquad H_2u_2 = \frac{1}{3\sqrt{2}}\,u_2$$

$$H_1u_3 = 0 \qquad H_2u_3 = -\frac{\sqrt{2}}{3}\,u_3$$

The eigenvalues can be considered the components of 3 two-component vectors called the *weight vectors,* or simply *weights*. The components of these vectors can be arrayed on a two-dimensional plot called a *weight diagram*. The three possible sets of components for the fundamental representation are $[1/\sqrt{6},1/(3\sqrt{2})]$, $[-1/\sqrt{6},1/(3\sqrt{2})]$, and $(0,-\sqrt{2}/3)$. Except for a scale factor, these correspond to the first triplet representation shown in Fig. 8-4. The $\bar{\mathbf{3}}$ triplet is obtained from a second fundamental representation whose weights are just the negatives of those above. The sextuplets and decuplets correspond to higher-dimensionality representations.

[1] These correspond to the three quarks in the application to elementary particles.

bibliography

This list is by no means exhaustive. More advanced treatments are purposely omitted. Most of the pre-1960 books on elementary particles are too out of date to be of much value and are not included.

GENERAL BIBLIOGRAPHY

BOOKS

Burhop, E. H. S., ed.: "High Energy Physics," Academic, New York, 1967. This series of five volumes contains articles on various topics. The level is advanced, but many of the articles include much useful material.

Benedetti, S. de: "Nuclear Interactions," Wiley, New York, 1964. This is a rather advanced text on nuclear physics.

Feld, B. T.: "Models of Elementary Particles," Blaisdell, Waltham, Mass., 1969. Basically a graduate-level text, but it covers in more detail much of the material of this book.

Frazer, W. R.: "Elementary Particles," Prentice-Hall, Englewood Cliffs, N.J., 1966. This is a graduate-level text but should be useful to the student who wishes to read further on the subject.

Frisch, D. H., and A. M. Thorndike: "Elementary Particles," Van Nostrand, Princeton, N.J., 1964. A very elementary book designed for the inquisitive layman.

Gouiran R.: "Particles and Accelerators," McGraw-Hill, New York, 1967. Elementary and rather philosophical; definitely loses much in translation.

Heckman, H. H., and P. W. Starring: "Nuclear Physics and the Elementary Particles," Holt, New York, 1963. This is an undergraduate-level text, mostly on nuclear physics, but it has useful material on elementary particles.

Levi Setti, R.: "Elementary Particles," University of Chicago Press, Chicago, 1963. A set of notes for an introductory graduate course. Rather advanced but useful.

Livingston, M. S.: "Particle Physics: The High-energy Frontier," McGraw-Hill, New York, 1968. Semipopular.

Lock, W. O.: "High Energy Nuclear Physics," Methuen, London, 1960. Readable, but somewhat outdated. Discusses elementary particles as related to nuclear physics.

Schiff, L. I.: "Quantum Mechanics," 3d ed., McGraw-Hill, New York, 1968. There are many good references on quantum mechanics. This is one of the better ones.

Segrè, E. G.: "Nuclei and Particles," Benjamin, New York, 1964. This is an excellent reference for nuclear physics and contains considerable material on elementary particles.

Swartz, C. E.: "The Fundamental Particles," Addison-Wesley, Reading, Mass., 1965. Semipopular.

Williams, W. S. C.: "An Introduction to Elementary Particles," 2d ed., Academic, New York, 1971. This book was intended as an introduction to theoretical methods. It is rather advanced but might be a good starting point for someone interested in a more theoretical approach than that of this text.

JOURNALS AND CONFERENCE PROCEEDINGS

Articles on current research are published in many journals. The most important of these are *Physical Review Letters* (*Phys. Rev. Letters*), *Physics Letters B* (*Phys. Letters*) and *The Physical Review D* (*Phys. Rev.*). Most of the articles are too advanced or abbreviated for the neophyte, but some acquaintance with the current literature is desirable. There are also numerous review articles in the *Annual Reviews of Nuclear Science* (*Ann. Rev. Nucl. Sci.*). These are advanced, but many of them are at a readable level. Brief, fairly elementary reviews also appear in *Comments on Nuclear and Particle Physics*. *Scientific American* (*Sci. Am.*) has many useful articles at a popular level. *Physics Today* (*Phys. Today*) also has many useful articles at a fairly elementary level. Articles, both published and unpublished, are abstracted and indexed in *Nuclear Science Abstracts*.

Many articles and reviews are also published in proceedings of numerous conferences on high-energy physics. The most important of these is the biennial "International Conference on High-energy Physics" (various publishers).

GOALS AND PROSPECTS OF HIGH ENERGY PHYSICS

"High Energy Physics Research: Hearings before the Subcommittee on Research, Development, and Radiation of the Joint Committee on Atomic Energy," GPO, Washington, 1965. This presents testimony by many leading high-energy physicists before a congressional committee discussing the aims, accomplishments, and needs of high-energy physics. There is considerable discussion on the need for higher-energy accelerators and other large related facilities. This volume provides a fascinating insight into the politics of basic physics research as seen by both physicist and politician.

CHAPTER BIBLIOGRAPHY

CHAP. 2 TECHNIQUES AND ACCELERATORS

Allkofer, O. C., W. D. Dau, and C. Grupen: "Spark Chambers," Verlag Karl Thiemig, Munich, 1969.

Blewett, J. P.: Resource Letter PA-1 on Particle Accelerators, *Am. J. Phys.* **34,** 1 (1966). Contains a brief history of particle accelerators and an extensive bibliography.

Livingston, M. S.: "High-Energy Accelerators," Interscience, New York, 1954. A detailed discussion of various types of accelerators with some theory of operation.

Livingston, M. S., ed.: "Development of High-Energy Accelerators," Dover, New York, 1966. A collection of articles relating to the development of accelerators.

Neal, R. B., ed.: "The Stanford Two-mile Accelerator," Benjamin, New York, 1968. A detailed description of the 20-GeV electron accelerator at the Stanford Linear Accelerator Center.

Ritson, D. M., ed.: "Techniques of High Energy Physics," Interscience, New York, 1961.
Shutt, R. P., ed.: "Bubble and Spark Chambers," Academic, New York, 1967.
Wenzel, W. A.: Spark Chambers, *Ann. Rev. Nucl. Sci.* **14,** 205 (1964).

There are also several journals devoted to instrumentation and accelerators. These are *Nuclear Instruments and Methods, Review of Scientific Instruments,* and *Particle Accelerators.* A very useful reference on instrumentation for high-energy physics is "Proceedings of the International Conferences on Instrumentation for High-energy Physics," biennial, various publishers.

CHAP. 3 STRONG INTERACTIONS

Most of the general references listed above contain considerable material on strong interactions.

Adair, R., and E. Fowler: "Strange Particles," Interscience-Wiley, New York, 1963.
Cence, R. J.: "Pion-Nucleon Scattering," Princeton, Princeton, N.J., 1969.
Chew, G. F., M. Gell-Mann, and A. H. Rosenfeld: Strongly Interacting Particles, *Sci. Am.* **210,** 74 (February, 1964).
Jeffries, C. D.: "Dynamic Nuclear Orientation," Interscience-Wiley, New York, 1963. A useful reference on polarized targets.
Moorhouse, R. G.: Pion-Nucleon Interactions, *Ann. Rev. Nucl. Sci.* **19,** 301 (1969).
Wilson, R.: "The Nucleon-Nucleon Interaction," Interscience-Wiley, New York, 1963. Compilation of data with description of experimental techniques and some theory.

CHAP. 4 CALCULATION OF RATES

Fermi, E.: "Elementary Particles," Yale, New Haven, Conn., 1951. A nice introduction to the application of field theory to elementary particles.
Feynman, R. P.: "Theory of Fundamental Processes," Benjamin,

New York, 1962. For those interested in a more rigorous and detailed treatment of the theory, this is a classic on calculational techniques. Advanced.

CHAP. 5 ELECTROMAGNETIC INTERACTIONS

Gatto, R.: Analysis of Present Evidence on the Validity of Quantum Electrodynamics, in E. H. S. Burhop, ed., "High Energy Physics," vol. II, Academic, New York, 1967.

Griffy, T. A., and L. I. Schiff: Electromagnetic Form Factors, in *idem*, vol. I.

Hofstadter, R.: "Nuclear and Nucleon Structure," Benjamin, New York, 1963. A reprint volume containing notable pre-1962 articles on electron scattering.

Hughes, E. B., T. A. Griffy, M. R. Yearian, and R. Hofstadter: *Phys. Rev.* **139,** B458 (1965).

Kendall, H. W., and W. K. H. Panofsky: The Structure of the Proton and the Neutron, *Sci. Am.* **224,** 60 (June, 1971).

Another very useful reference source is "Proceedings of the International Symposiums on Electron and Photon Interactions at High Energy," biennial, various publishers.

CHAPS. 6 AND 7 WEAK INTERACTIONS; THE K^0–$\bar{K}^0$ SYSTEM

Feinburg, G., and L. M. Lederman: The Physics of Muons and Muon Neutrinos, *Ann. Rev. Nucl. Sci.* **13,** 431 (1963).

Kabir, P. K.: "The CP Puzzle," Academic, New York, 1968. A lucid discussion of the experimental data regarding CP violation and its theoretical interpretation.

Lederman, L. M.: Neutrino Physics, in E. H. S. Burhop, ed., "High Energy Physics," vol. II, Academic, New York, 1967.

Lee, T. D. and C. S. Wu: Weak Interactions, *Ann. Rev. Nucl. Sci.* **15,** 381 (1965); **16,** 471 (1966); **16,** 511 (1966).

Marshak, R. E., Riazuddin, and C. P. Ryan: "Theory of Weak Interactions in Particle Physics," Interscience-Wiley, New York, 1969. An advanced but very complete and up-to-date discussion of weak interactions.

Rubbia, C.: Weak Interaction Physics, in Burhop, *op. cit.*, vol. III.

CHAP. 8 THE EIGHTFOLD WAY, QUARKS, AND SU(3) SYMMETRY

Brown, L. M.: Quarkways to Particle Symmetry, *Phys. Today* **19,** 44 (February, 1966).

Hamermesh, M.: "Group Theory and Its Application to Physical Problems," Addison-Wesley, Reading, Mass., 1962.

Lichtenberg, D. B.: "Unitary Symmetry and the Elementary Particles," Academic, New York, 1970.

Lipkin, H. J.: "Lie Groups for Pedestrians," North-Holland, Amsterdam, 1965.

Matthews, P. T.: Unitary Symmetry, in E. H. S. Burhop, ed., "High Energy Physics," vol. I, Academic, New York, 1967.

Rowlatt, P.: "Group Theory and the Elementary Particles," Longmans, London, 1966.

Tuan, S. F.: *Phys. Today* **21,** 31 (January, 1968).

Zeldovich, Ya. B.: Classification of Elementary Particles and Quarks "for the Layman," *Sov. Phys. Usp.* **8,** 489 (1965).

INDEX

INDEX